Created in cooperation with Fluke Corporation

Energy Auditing for Industrial Facilities

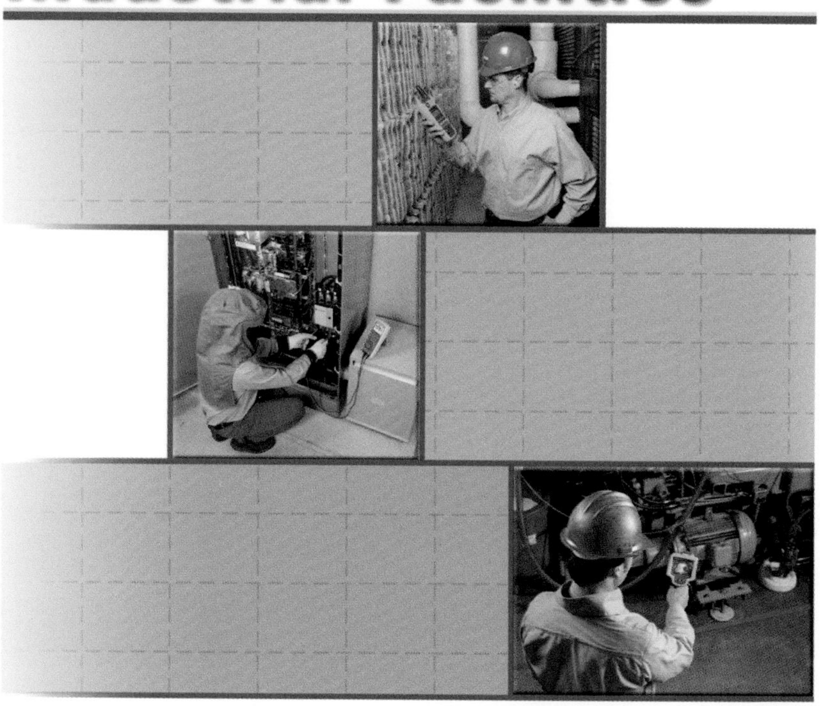

AMERICAN TECHNICAL PUBLISHERS
ORLAND PARK, ILLINOIS 60467-5756

Acknowledgments
Cleaver-Brooks
Fluke Corporation
Mine Safety Appliances Co.

1 2 3 4 5 6 7 8 9 – 11 – 9 8 7 6 5 4 3 2 1

Printed in the United States of America

ISBN 978-0-8269-1521-4

This book is printed on recycled paper.

Contents

CD-ROM Contents

- Quick Quizzes®
- Illustrated Glossary
- Flash Cards
- Fluke Virtual Meters
- Reports and Forms
- Media Clips
- ATPeResources.com

Introduction

Improving energy efficiency is important to managing energy use, which reduces utility and maintenance costs and provides additional benefits. As significant energy consumers, industrial facilities have much to gain from improving energy efficiency.

Opportunities for improving energy efficiency in existing facilities are identified and quantified through an energy auditing process. During an energy audit, each building system is investigated for potential causes of energy or resource waste, and projects to reduce energy use are then planned.

Energy Auditing for Industrial Facilities is an introduction to planning, conducting, documenting, and evaluating an energy audit in an industrial setting. Specific energy audit measurements are suggested for each system type, along with the significance of the results. The book describes how to prioritize, implement, and verify the recommended projects to improve efficiency and includes suggestions for sustaining the results.

After studying this material, technicians familiar with building systems and the use of test instruments will be better prepared to participate in an energy audit. This book is also important for educating managers on the benefits and process of conducting an energy audit.

The CD-ROM included in the back of the book includes several electronic resources for reinforcing the technical content, including Quick Quizzes®, an Illustrated Glossary, Flash Cards, Fluke Virtual Meters, Reports and Forms, Media Clips, and a link to additional content at ATPeResources.com. Information about using the *Energy Auditing for Industrial Facilities* Interactive CD-ROM is included with the book.

The Publisher

chapter 1

Energy Audits

An energy audit is an investigation into ways to improve the energy efficiency of a facility. The audit and its recommended changes usually require an investment into the replacement and repair of building systems and equipment. However, this provides savings in the form of reduced utility costs that pay back the facility owner within months or a few years.

ENERGY AUDITING

An *energy audit* is a comprehensive review of the energy use of a facility that results in a report regarding ways to reduce energy use through changes to buildings, equipment, and procedures. The audit collects detailed information about every use of a particular utility resource, such as electricity, and prioritizes strategies to reduce the consumption.

Recommended changes may include improving building insulation, replacing equipment with more energy-efficient models, implementing more sophisticated system controls, changing maintenance procedures, repairing leaks, and changing occupant activities. The primary goal is the reduction of operating costs through lower energy use without negatively impacting facility operations, processes, occupant comfort, or safety.

Return on Investment

Some energy savings can be achieved by only changing consumption habits, which have little or no costs and bring immediate results. However, many energy efficiency measures, and the audit itself, incur financial costs and are usually implemented over a period of time. The goal of an audit is to identify the measures that result in savings over a reasonable period that outweigh the cost of the implementation.

Return on investment (ROI) is the ratio of a project's financial gain to its cost. Since savings (and sometimes costs) are often periodic, ROI may be referenced to a length of time. *Payback period* is the time elapsed until a project's resulting savings equals its costs. **See Figure 1-1.** For example, replacing a certain piece of equipment with a version that is more energy efficient will save about $1000 per year in energy costs. If the new equipment has a one-time cost of $2000, then the payback period is two years. After three years, the savings totals $3000, which is a net gain of $1000 and an ROI of 150% ($3000 ÷ $2000 = 150%). In four years, the ROI is 200%. Both ROI and payback periods are measures of cost effectiveness.

The types and magnitudes of potential savings opportunities vary greatly between facilities. Generally, facilities with older and larger equipment realize larger savings and shorter payback periods. Alternatively, facilities with proactive maintenance programs that use preventive and predictive maintenance techniques have already accomplished many audit goals. These facilities will likely find few additional savings opportunities.

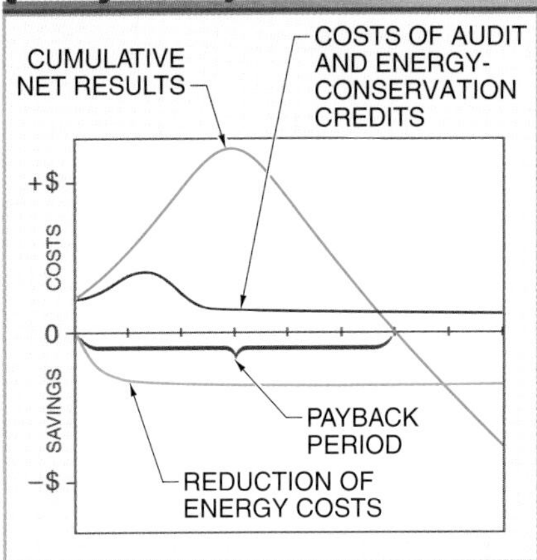

Figure 1-1. A return on investment (ROI) analysis compares an initial expenditure with the resulting savings to determine the cost effectiveness of a plan.

Energy Efficiency Benefits

The primary goal of performing an energy audit is the identification of ways to improve energy efficiency (using less energy to provide the same or a better result). The most obvious benefit of this is reduced utility costs. However, there are often several other benefits, such as additional financial, safety, morale, or public relations advantages.

Reduced Utility Costs. Improved energy efficiency reduces overall energy use, which results in lower utility costs. These results should be directly reflected in the utility bills for the months following the implementation of efficiency-related changes.

Moreover, a reduction in electricity use may lower costs in multiple ways. **See Figure 1-2.** First, at a certain electricity rate, as consumption falls, so does the total cost of electricity use. However, for large electricity consumers, such as commercial and industrial facilities, the rate may vary depending on the power demand. If the peak power demand is reduced sufficiently, the rate is lowered, reducing costs further. Finally, improving the power quality of the electrical system increases the efficiency of purchased electricity. This also reduces power demand and consumption.

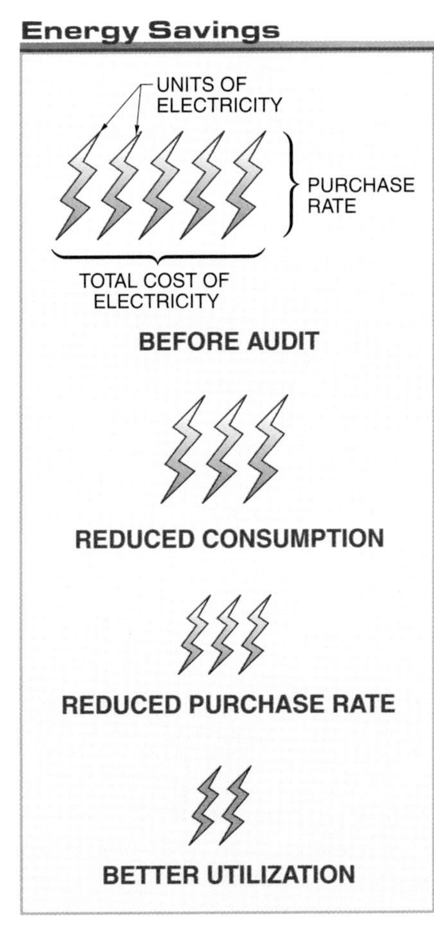

Figure 1-2. Improving the efficiency of electricity use can provide financial savings in multiple ways. Benefits include reduced overall consumption, reduced rates due to reduced peak power demand, and better utilization of purchased power.

Financially, the reduction in utility costs has several benefits. For example, operating costs are lowered, which results in higher margins and increased profitability. Cash flow also improves, and working capital is available for other projects. Energy prices will continue to rise in the long term, so carefully managing these costs improves business stability.

Improved Occupant Safety and Comfort. Lighting and HVAC systems are particularly important systems in energy audits because they are typically the largest energy consumers. They are also important to the safety and comfort of building occupants. Therefore, improvements to these systems often improve the working environment as well. Also, the repairing of leaks in compressed air, steam, and water systems reduce potential safety hazards.

Reduced Maintenance and Downtime Costs. Measures taken to improve energy and resource efficiency typically involve replacing and overhauling old equipment. Newer or properly repaired equipment is less likely to require breakdown maintenance, which is costly emergency work. Also, if equipment breakdowns are less likely, the resulting production downtime is reduced.

Interestingly, an energy audit report may recommend increasing preventive maintenance activities in order to preserve the high efficiency operation of new equipment. However, these regular maintenance tasks, such as cleaning, lubrication, and inspection, are far less expensive than breakdown maintenance and can usually be scheduled for convenient times.

Incentives and Rebates. The federal government, along with many state governments and local municipalities, offer financial incentives to help offset the costs of energy efficiency improvements. These incentives are typically offered as various programs to reduce tax liability, such as tax credits, deductions, and rebates. All programs require some type of documentation or certification as proof of the implementation and verify the expected results. Therefore, in order to take advantage of one or more of these programs, it is likely necessary to plan for the requirements early, during the audit phase.

Green Building Certification. To help promote energy-efficient facility operations, some organizations have developed rating systems and achievement levels to award building owners and operators for environmentally friendly practices. One of the most prominent programs is from the U.S. Green Building Council®. Their Leadership in Energy and Environmental Design® (LEED®) Green Building Rating System™ is the nationally accepted standard for the design, construction, and operation of green buildings. **See Figure 1-3.**

The LEED certification program is focused primarily on commercial buildings but is applied to industrial facilities as well. Most certification categories are intended for new construction, but there is a special category for existing buildings that concentrates on the ongoing operations and maintenance of a building.

TECH-TIP

Most industrial facilities can reduce their utility expenses by about 15% to 20% with minimal investment, and some by up to 30% with equipment upgrades. Payback periods are typically at least a few months, but should be less than two years.

LEED® Green Building Rating System™

Key Measurement Areas	Rating Categories
• Sustainable Sites	• New Construction
• Water Efficiency	• Existing Buildings: Operations & Maintenance
• Energy & Atmosphere	• Commercial Interiors
• Materials & Resources	• Core & Shell
• Indoor Environmental Quality	• Schools
• Location & Linkages	• Retail
• Awareness & Education	• Healthcare
• Innovation in Design	• Homes
• Regional Priority	• Neighborhood Development

Figure 1-3. The LEED® Green Building Rating System™ rewards the efficient use of energy and resources with certifications based on achievements in several key areas. Green building certifications boost internal morale and external public relations.

ENERGY AUDIT TYPES

Energy auditing includes a broad spectrum of studies that ranges from a quick walk-through to a comprehensive analysis of all energy consumption. There are three common types of audit programs. The differences between them are the level of breadth and detail involved in the investigation. These factors affect the cost of the audit, the accuracy of the resulting analysis, and the possible solutions presented.

The costs associated with conducting an audit include employee time for planning and conducting the audit, the procurement of necessary test instruments, the hiring of outside consultants, and lost production if processes must be temporarily shut down for measurements. These costs vary greatly depending on the facility and the scope and type of audit.

As an audit becomes more thorough, more opportunities for savings are typically uncovered. However, the savings opportunities are often of progressively smaller benefit. **See Figure 1-4.** For example, a simple audit that costs $10,000 may uncover opportunities for saving $20,000 per year. Increasing the scope of the audit to $50,000, however, may yield only another $20,000 in potential savings. Of course, the costs of implementing the changes recommended by the audit must also be considered.

Therefore, it is necessary to evaluate the potential efficiency gains against the growing audit-related costs at multiple times during the investigation in order to decide whether to go further. Each facility has a different situation, so the most cost-effective solution will vary.

Preliminary Energy Audits

A *preliminary energy audit* is an overview of the major energy-consuming processes of a facility to identify only significant inefficiencies. This type of audit is also called a simple audit, screening audit, or walk-through audit. This is the simplest and quickest type of audit but often provides limited benefit. It can only uncover the most obvious energy waste or inefficiency problems.

Energy Audit Cost Effectiveness

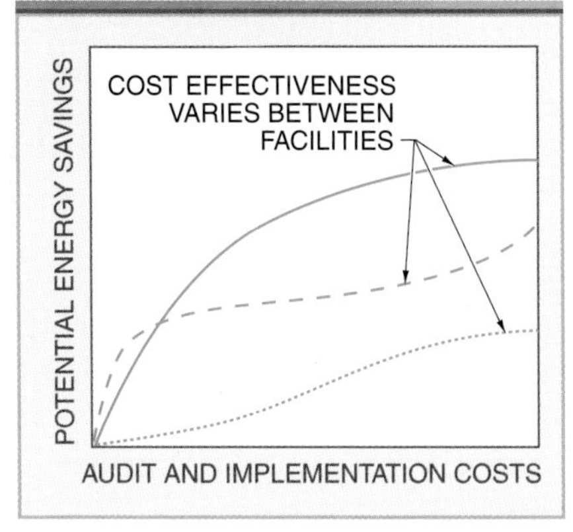

Figure 1-4. Increasing the scope and therefore, expense of an energy audit typically yields additional opportunities for energy savings, but this is not necessarily cost effective. The appropriate size and type of audit vary between facilities.

A preliminary energy audit involves a walk-through of the facility to become familiar with the building operation, interviews with facility personnel, and a review of utility bills and other operating data. This brief audit may not include any test instrument measurements. The result includes a simple prioritized list of corrective measures, cost estimates, and potential savings.

Typically, only major problems, such as significant leaks or inappropriate setpoints, are uncovered during this type of audit. However, depending on the situation, this may still be the most cost-effective option. Based on the results, it is then determined whether a more detailed audit is appropriate.

General Energy Audits

A *general energy audit* is a facility energy study that expands on the preliminary energy audit and requires collecting more information and performing a more detailed analysis. This type of audit requires a record of utility bills over a 12-month to 36-month period. The record reveals the short- and long-term consumption patterns and rate changes.

Current data for energy and resource usage per system or equipment is tracked by extensively measuring system parameters with test instruments. The data and other details are recorded on audit forms for later analysis.

This type of audit identifies and prioritizes appropriate energy conservation and efficiency measures. Also included are the implementation cost estimates, site-specific operating cost savings opportunities, and customer investment criteria. The higher level of detail in a general energy audit is often needed to justify the significant implementation costs.

Investment-Grade Energy Audits

In large corporate facilities, the types of changes recommended as a result of energy audits are often major investments that require executive scrutiny and financial analysis. The decision-making process is based on the expected ROI. To provide the completeness and accuracy required for the analysis, a facility may conduct an investment-grade energy audit.

Fluke Corporation

The majority of energy audit tasks involve taking measurements, although planning, research, analysis, and report generation are also critical parts of the audit process.

An *investment-grade energy audit* is a comprehensive facility energy study that expands on the general energy audit and involves developing a dynamic model of the energy-use characteristics of a facility. The model is calibrated against actual utility data to ensure that it provides a realistic baseline. Then, the model is used to simulate the results of proposed energy conservation and efficiency measures. This provides the most accurate estimates of potential energy savings, which are then included in the financial analysis reports. The extensive data gathering and model testing of an investment-grade energy audit earns a high level of confidence in the conclusions of the final reports.

chapter 2

Energy Audit Process

An energy audit is a four-step process that includes the phases of planning, investigating, implementing, and sustaining. Thorough planning and careful action increase the chances of a successful energy audit with a maximum return on investment.

AUDIT PROCESS

An energy audit includes more than just taking measurements. In order for the measurements to be useful, they must be part of a systematic procedure to identify and implement the most cost-effective energy-conservation programs. Energy audits involve gathering system information, measuring energy use, developing conservation strategies, choosing the most cost-effective plan, implementing changes, and verifying results. **See Figure 2-1.** Also, a proactive facility continues improving energy efficiency by repeating this process at regular intervals.

Audit Process

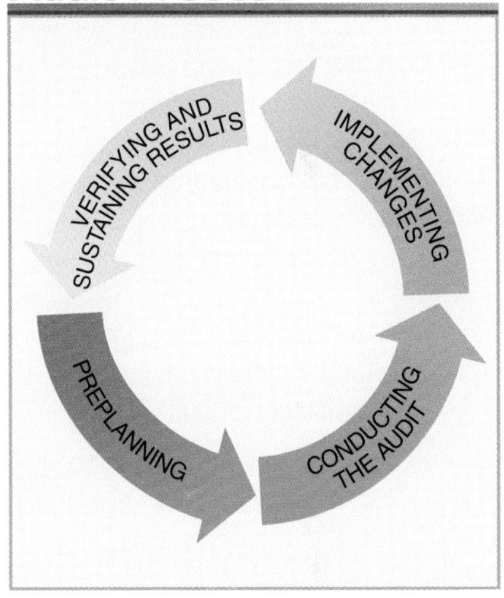

Figure 2-1. The audit process consists of four phases. The audit process is repeated at intervals to continually improve energy efficiency.

Preplanning

Energy audit preplanning begins with making a commitment to energy conservation. Multiple levels of a business must be involved and dedicated to making the entire audit process a priority. The initial goals of the preplanning phase are to establish the audit team members, decide on the scope of the audit, develop a timeline of tasks to be completed, and assign team member responsibilities. The preplanning process typically lasts between a few weeks and a few months, depending on the size of the facility and scope of the audit.

Audit Teams. An audit team is formed from the personnel of different departments. A relatively small group, such as three to five employees, is tasked with the bulk of audit work. The group usually consists of maintenance personnel, who are the most familiar with the building systems and equipment. **See Figure 2-2.** However, other team members coordinate certain supporting tasks. Accountants gather and organize data from utility bills. Production staffers coordinate audit activities with employee and operating schedules. High-level managers are closely involved so that necessary decisions can be made quickly in order to facilitate audit tasks.

Audit Scope. The audit team must decide on the initial scope of the audit. This includes which systems will be investigated and the depth of the investigations. Facility maintenance personnel may already suspect where significant energy wastes are and what tests will be required to quantify them. This information is usually the basis for deciding on the type of audit. Audit scope can always be expanded later if initial results lead to deeper testing. However, it is helpful to establish early the criteria for allowing the scope to be expanded along with how other phases of the process will be affected.

Baseline Data Gathering. An energy audit relies heavily on comparisons between expected energy use and actual use. Actual energy use is quantified as baseline and measured data. *Baseline data* is data that represents a normal operating state and is used as a point of reference for future changes. Most measured data is gathered later during an audit investigation, but baseline data is gathered primarily from existing documents. The most common sources of baseline energy use data are utility bills. Billing information should cover at least the preceding 12 months. However, gathering information from over a few years may better represent any seasonal or long-term trends.

Other documentation that is useful when planning an energy audit or analyzing results includes equipment specifications, manufacturer recommendations for testing equipment, and maintenance records. Audit team members may also want to interview personnel who work on or near major equipment about their observations and suggestions.

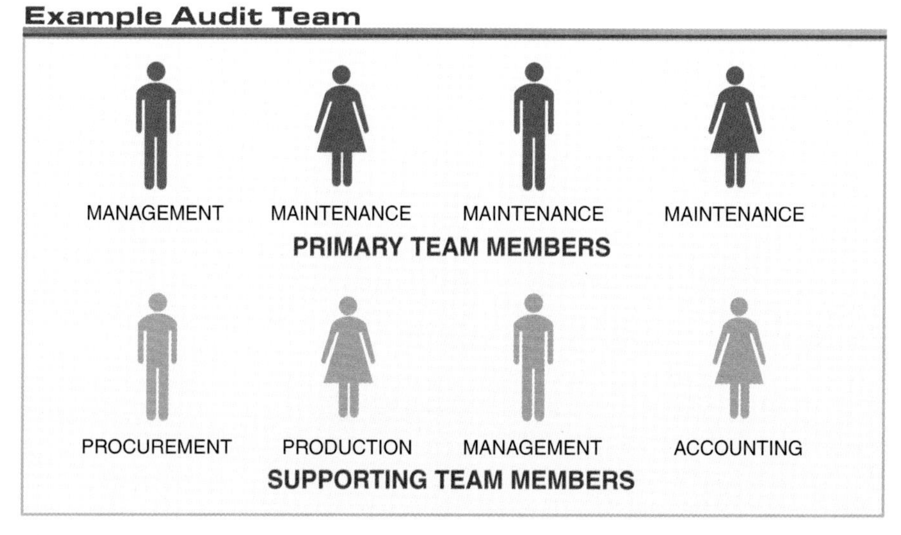

Figure 2-2. A few individuals form the primary part of the audit team, but they receive support and assistance from other personnel, particularly those in other departments.

Plant Profile. Preliminary information is used to create a plant profile. This is a snapshot view of the facility including square footage, energy expenditures, significant loads, control system setpoints, and other basic facts.

Conducting the Energy Audit

The auditing portion of an energy audit consists of two phases. First, the core members of the audit team conduct an investigation and then help prepare an audit report. The supporting members of the audit team are also involved in preparing the audit report. Outside contractors may be involved in any phase for their expertise or specialized tools.

Audit Investigations. An audit investigation involves the inspection of each system within the scope of the audit. It traces from the source of energy or the resource to each point of use. A variety of test instruments is used extensively during the investigation to identify energy waste and other abnormal conditions. **See Figure 2-3.** The audit assesses the efficiency, physical condition, and operating profile of the equipment, including the duty cycle, load changes, and controls.

Resources are available from equipment or test instrument manufacturers that provide guidelines of how to take measurements and recognize potential energy waste issues. The length of this phase may last from several days to several weeks, depending on the size of the facility and whether extended data logging is needed. Collected data is recorded for later analysis.

Audit Reports. The second phase of an audit is the preparation of the audit report. On the report, the completed investigation is summarized and the findings are presented in an organized and prioritized format.

Test Instruments

Fluke Corporation

Figure 2-3. Test instruments are used extensively throughout the baseline data gathering and audit investigation phases.

The financial cost of each instance of wasted energy is calculated, based on the audit measurements. Related multiple issues may be grouped together for analysis. For example, a series of three leaks in a compressed air system is causing the compressor to operate approximately 50% more than expected. The cost of the extra electricity consumed is estimated from the electrical measurements of the audit and the rate data from the billing information. This is the amount that would be saved if the leaks were eliminated.

For each energy problem, one or more possible remedies are listed, along with their estimated total costs. **See Figure 2-4.** Total costs must include any financial impact of completing the project, including equipment expenses, installation or repair labor, employee training, and process downtime. The financial analysis portion presents the ROI estimates for each option. For example, the repair of the compressed air distribution system to eliminate the leaks would incur a certain cost in parts and labor, which is compared to the estimated cost savings to determine the payback period.

Audit Report Analysis

Problem	Cost of Energy Loss	
Compressed air distribution system has seven leaks	$560/month	
Possible Solutions	**Estimated Project Costs**	**Payback**
Repair leaks in existing distribution system	$800	1.5 months
Replace all piping to eliminate unnecessary connections and reroute around areas where piping is prone to physical damage	$3,600	6.5 months

Figure 2-4. An audit report lists all uncovered sources of energy waste, along with possible solutions and the estimated payback.

Each project is usually listed in the report in order of payback period, unless there are other factors that affect its importance, such as safety. This prioritization provides a simple way for decision makers to evaluate the relative cost-effectiveness of the recommendations.

Implementing Changes

Based on the results outlined in the audit report, decision makers evaluate the recommended projects against budget and other constraints. Some or all of the recommendations are approved, and an action plan is developed to facilitate the implementation. Team members are assigned responsibilities to initiate and monitor each project, ensure that it is completed according to the recommendation, and consult with the audit team if changes to the project plan become necessary. Project planning should also include preparation for how the success of the project will be measured.

Verifying and Sustaining Results

As energy conservation projects are completed, the results should be measured to determine whether the goals of the project were achieved. Audit-type measurements should be conducted again to verify the expected energy savings and confirm that there were no negative effects. It is highly recommended to install permanent monitoring equipment on the largest or most critical loads to continuously measure energy consumption.

Maintenance programs should be adjusted as needed to help sustain the energy savings. This typically involves improving preventive and predictive maintenance activities, along with supporting prompt and effective troubleshooting and repairs in the event of a failure.

Complete follow-up audits should be conducted at periodic intervals to verify expected energy use and identify long-term trends toward inefficiency. Simple walk-through audits may be conducted monthly or quarterly and more thorough audits conducted at longer intervals. The audit process then begins again with preplanning, forming a continuous cycle of efficiency improvement.

AUDIT SCOPE

An energy audit may include every energy-using system in the facility. However, if it is not feasible to conduct such a comprehensive audit, or if the potential savings are not expected to justify the expense, the energy audit is targeted at areas that are expected to yield the largest improvements. Some systems are likely to have better ROI than others. **See Figure 2-5.** Typically, these are the HVAC and lighting systems.

Typical ROI for Building Systems

High	HVAC Lighting
Moderate	Electrical power quality Electrical motors and drives Compressed air
Low	Building envelope Plug-in loads Steam systems Water systems Waste and recycling

Figure 2-5. The initial scope of an energy audit should include the systems that are most likely to have a high ROI, meaning that significant energy wastes can be remedied with relatively small investments.

In industrial facilities, energy is used to operate both the processes and the building. A *process* is a series of actions intended to produce, assemble, or modify an industrial product. For example, an assembly line is a process where component parts are put together to form a final product. Other industrial processes include refining raw materials, cooking food products, and blending chemicals. The process equipment uses energy and resources to operate equipment, move materials, and perform any other actions needed to transform the starting materials to the desired product.

During an energy audit, often systems are investigated that supply a resource to a process, such as steam. However, the energy efficiency of the process itself is typically beyond the scope of the building audit. If desired, an energy audit of the process is conducted separately. For example, the entire steam system between the boiler and the process may be within the building audit scope but not the process equipment that uses the steam at the end point.

AUDIT RESOURCES

There are a number of resources to help businesses and homeowners conduct energy audits and implement energy efficiency projects. A number of guides, forms, templates, spreadsheets, and software tools are available for organizing information, performing calculations, and formatting reports. Many of these are free and available online.

TECH-TIP

An energy audit surveys any type of energy use, including electricity, steam, compressed air, chilled water, and fossil fuels. Other resources, such as water, may also be surveyed for savings opportunities.

Comprehensive energy-efficiency information is provided by the ENERGY STAR® program. This is a joint program between the U.S. Environmental Protection Agency and the U.S. Department of Energy. ENERGY STAR recognition is applied to products that meet minimum energy-efficiency criteria. ENERGY STAR also provides technical information and tools that businesses can use to conduct energy audits and implement energy efficiency solutions. Data collected from national surveys allows facility owners to compare energy use with other similar facilities nationwide.

chapter 3

Auditing HVAC Systems

Reducing HVAC system energy use does not involve simply turning all equipment to a lower setting. This type of action could create indoor air quality problems and health hazards, and the savings may not even be significant. Rather, smart choices in changing certain HVAC equipment and/or the way it operates can greatly improve system efficiency while maintaining or improving occupant comfort.

HVAC SYSTEMS

A *heating, ventilating, and air conditioning (HVAC) system* is a system used to condition air by maintaining proper temperature, humidity, and air quality. Properly conditioned air improves the comfort and health of building occupants. Excessive airborne dirt, heat, and humidity can cause irritation and injury to people and possible damage to sensitive equipment. HVAC systems are designed to operate at optimum energy efficiency while maintaining desired environmental conditions.

Temperature Control

Indoor temperature is controlled by transferring heat to or from the indoor air with heat exchangers. A *heat exchanger* is a device that transfers heat from one fluid to another fluid without allowing the fluids to mix. Most types of heat exchangers, also known as coils, are installed in HVAC ducts, where air is blown over the exchangers by fans. **See Figure 3-1.** Some types of heat exchangers are installed directly within the occupied spaces and rely on radiant heat and natural convection airflow to transfer heat.

Heat Exchangers

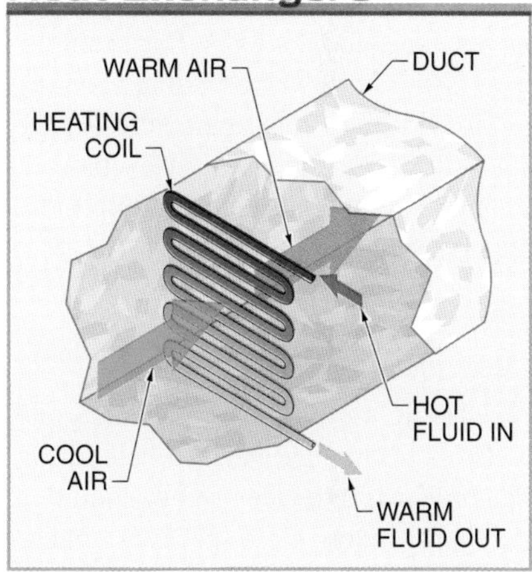

Figure 3-1. Heat exchangers allow the transfer of heat between two fluids, such as steam and air, without mixing.

Heating is provided by furnaces, boilers, or electric heaters. Furnaces burn a fuel, such as natural gas, and the hot combustion gases flow through a forced-air heat exchanger. Boilers supply hot water or steam that flows through heat exchangers, either forced-air or radiant types. Electric heaters produce heat from electric current flowing through resistive elements. Electric heaters are used in forced-air or radiant systems.

Cooling is provided by a refrigeration system, which either cools the indoor air directly or chills water that is then used to cool air. Cooling coil heat exchangers are used only in forced-air systems. Heat in the indoor air transfers to the cooling coil fluid, cooling the air.

The amount of heating or cooling is controlled by either regulating the temperature of the heat exchanger or the volume of airflow blown over the exchanger. As the indoor temperature changes, controllers automatically adjust valves, fans, and dampers. For example, opening the steam valve feeding a heating coil from 50% to 75% open increases the steam flow through the heat exchanger, raising its temperature and adding more heat to the indoor air.

Air Handlers

Most HVAC systems use air handlers and ductwork to circulate and distribute air. An *air handler* is a device used to condition and supply indoor air to spaces in a building. Air handlers include fans, dampers, heating coils, cooling coils, and humidifiers that are controlled to provide the desired air conditions. Sensors measuring temperature, humidity, pressure, airflow, and other characteristics monitor the air and provide input signals to controllers.

The air handler mixes outside air with return air from inside the building. **See Figure 3-2.** *Outside air* is fresh air from outside a building that is incorporated into an HVAC system. *Return air* is air from within a building space that is drawn back into an HVAC system to be exhausted or reconditioned. *Exhaust air* is air that is ejected from an HVAC system to the outside environment.

Mixed air is a blend of return air and outside air inside an air handler that goes on to be conditioned. Mixing outside and return air ensures that enough fresh air enters the building to prevent the air from becoming stagnant, while retaining some of the conditioning that has already been added to the indoor air. Incorporating outside air is also a way to obtain as much natural cooling as possible. Mixed air is filtered and conditioned to become supply air. *Supply air* is newly conditioned air that is distributed through supply ducts to building spaces.

Building Zones

Different areas within a facility may have different HVAC requirements. Therefore, the supply air to each area may need to be conditioned differently. A *zone* is an area within a building defined by its specific HVAC requirements. Zones may be defined by a desired climate condition and/or how precisely a condition must be controlled.

A building may have many zones, usually divided by the use of the space and to create manageable units of HVAC conditioned areas. For example, a facility with a mixture of office and assembly areas may have the HVAC system functions divided into zones by these different functions.

Some HVAC systems include all heating and cooling coils in the central air handler and distribute air that is conditioned to the aggregate requirements of all the areas. Dampers at each area only control how much of this conditioned air is admitted, affecting the indoor climate of each space slightly.

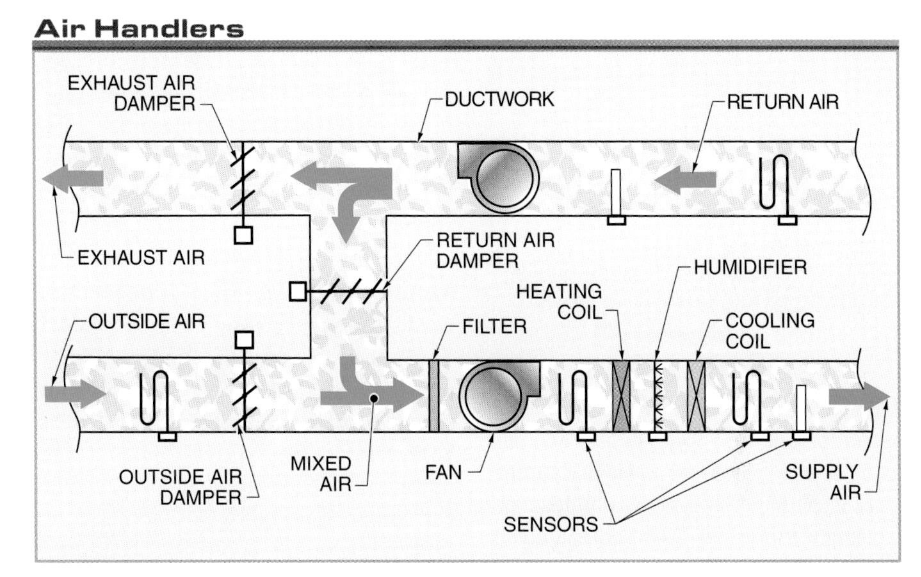

Figure 3-2. Air handlers are the primary units in most HVAC systems for conditioning and circulating air.

In multizone systems, the mixing of outside air and the filtration of the supply air occurs at one or more air handlers, but other functions, primarily heating and/or cooling, may be located downstream at terminal units. There are many HVAC system arrangements that can control the supply air temperature to multiple zones individually. The primary differences are the locations of the heating and cooling coils and how supply air is mixed.

Terminal Units

After being distributed from central air handlers, supply air enters the building space through a terminal unit. **See Figure 3-3.** This air's temperature is set to a general setpoint. A *terminal unit* is a device that is located close to an HVAC zone and heats or cools preconditioned supply air flowing through it to meet the particular requirements of the zone. Forced-air terminal units include fans or dampers to modulate the amount of conditioned supply air into the space and may include other devices to further condition the supply air.

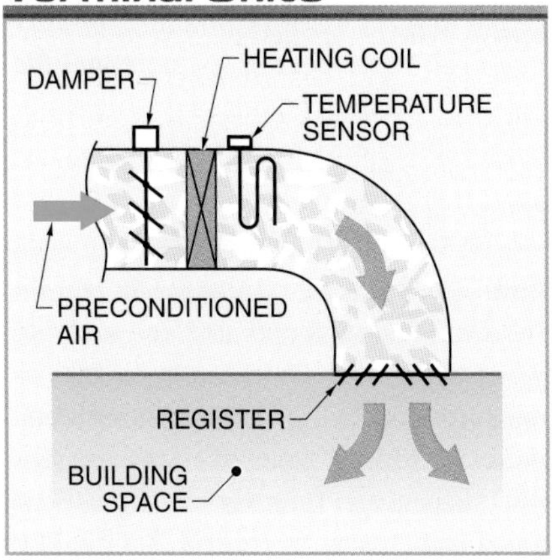

Figure 3-3. Terminal units are used if air must be further conditioned near a building space.

HVAC SYSTEM ENERGY LOSSES

HVAC systems can account for 40% to 60% of the energy consumption of a building. Inefficiencies and poor operations not only waste energy but may also adversely affect occupant health and productivity. Optimizing system performance involves identifying and remedying operating inefficiencies and sources of conditioned air losses.

Inefficient HVAC Equipment

As equipment ages, it gradually becomes less efficient due to wear and tear. Extreme service, damage, and poor maintenance accelerate this trend. Also, compared to newer equivalent models, the efficiency of even a well-maintained but old unit is often quite low.

Audit tasks measure the useful output (airflow, heat, pressure, etc.) of each unit against its input (usually electricity) and compare this to both its original specifications and the specifications of possible replacements. Some efficiency may be regained through repairs, but replacements are often the most effective option. Replacing HVAC equipment typically requires a significant investment, but the potential efficiency gain may still result in a reasonable ROI.

Control System Problems

Control systems determine when certain HVAC units should operate, for how long, and at what level, based on the information provided by sensor measurements. There are several types of HVAC system controls, and each can vary in sophistication. Problems with controls can easily reduce HVAC system efficiency by causing equipment to run unnecessarily.

Sensor Errors. System controls rely on sensors to accurately measure present conditions, both in the occupied areas and inside of air handlers and terminal units. If a sensor does not accurately represent the condition of the air, then it may be causing the HVAC system to over- or undercondition the supply air.

Motor Operation. Electric motors are used in many different types of HVAC equipment to operate compressors, pumps, and fans. Some controls can run motors at only 100% full speed or completely OFF, which causes large power draws and may wear out mechanical components quickly. Alternatively, improving motor controls and adding variable-frequency drives optimize motor operation and can reduce electricity and maintenance costs.

Unsuitable Setbacks. A *setback* is a temperature setpoint that is used when an area is unoccupied. A setback reduces the heating or cooling load, allowing the HVAC system to operate less when the building is empty. It also maintains a temperature that can be easily returned to the occupied setting. **See Figure 3-4.** For example, a heating setback may reduce the temperature setpoint from 70°F (when occupied) to 65°F (when unoccupied). Optimal setbacks vary by geographic region and building characteristics. If not set up properly, setbacks can actually reduce system efficiency.

TECH-TIP

HVAC controls should operate equipment in a way that maintains a comfortable indoor climate while minimizing energy use during all times of the year.

Setbacks

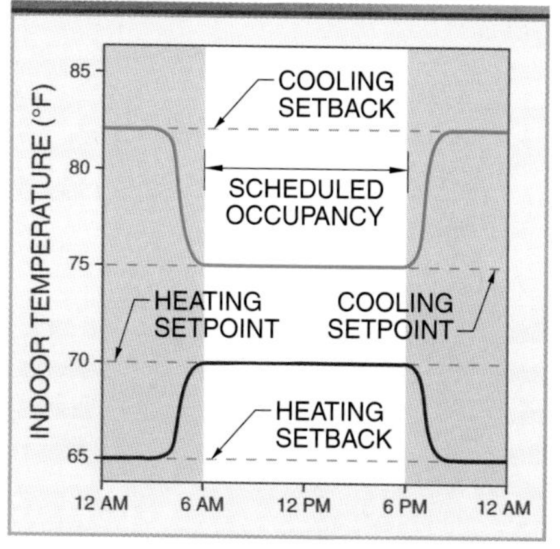

Figure 3-4. Setbacks reduce the heating or cooling load on an HVAC system during unoccupied hours, but they allow the system to return to occupied setpoints quickly.

Building Envelope Leaks

The indoor climate is contained by the building envelope. A *building envelope* is the continuous surface of walls, roof, floors, doors, and windows of a building that separates the building interior from the outside environment. Any escape of indoor conditioned air, or entry of unconditioned outside air, through the building envelope requires the HVAC system to work harder to maintain the desired conditions. This increases energy consumption. Also, if contaminated outside air enters the building, it may create a health hazard for the occupants.

Air leaks in a building envelope are common around poorly sealed doors or windows and transitions between different building materials. Loading docks, common in industrial facilities, are a major source of air leaks. Some loading docks use insulated and tight-fitting doors with a perimeter of compressible foam to seal the edge of the trailer door. However, loading docks are still particularly prone to leaking air due to worn out or damaged seals, misaligned trailers, and frequent door opening.

Building envelope areas that are sealed (preventing the actual exchange of indoor and outside air) but lacking adequate insulation are also a problem. Insulation prevents the transfer of heat through the building materials, in either direction. Therefore, poorly insulated areas cause the HVAC system to operate more in order to maintain the desired indoor temperature.

Extra Heat Sources

Many industrial processes involve heat, which is either added to the process or generated by the process itself, such as through chemical reactions. This extra heat, if not adequately sealed and insulated from the rest of the indoor space, adds another cooling load to the HVAC system.

Wasteful Ventilation

Air within a building gradually becomes stale from exhaled carbon dioxide, which is unhealthy for occupants. Indoor air is kept comfortable by diluting stale return air with fresh outside air. **See Figure 3-5.** *Ventilation* is the process of introducing fresh outside air into a building. Two sets of dampers control the relative proportions of outside air and return air, which are mixed to become supply air.

Minimum ventilation requirements are based on the building occupancy and type of indoor activity, such as office work, classes, or manufacturing. Typically, 5% to 30% of supply air is composed of outside air. Due to air contaminants, ventilation requirements for many industrial facilities are higher.

Wasteful Ventilation

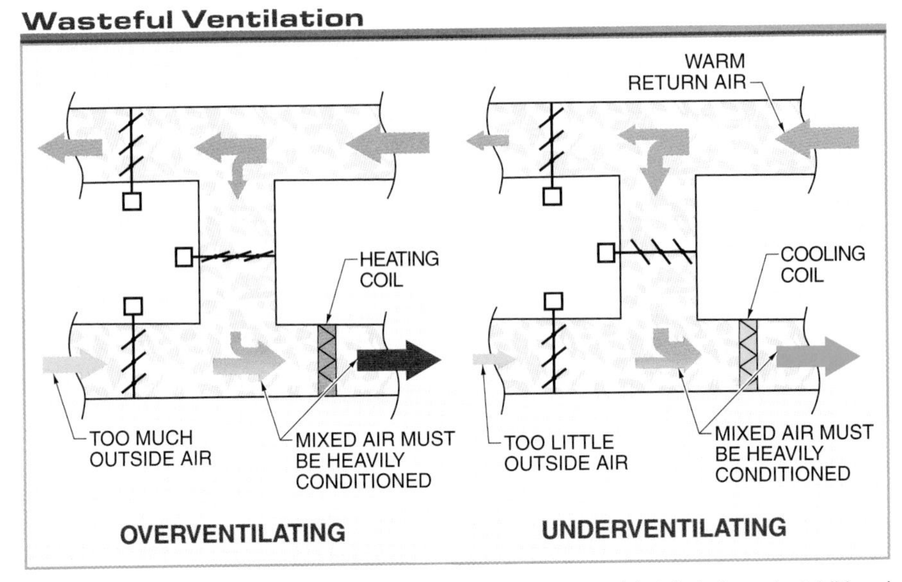

Figure 3-5. A minimum amount of ventilation is required to avoid stale indoor air. Additional ventilation may help reduce cooling loads but only under the right conditions.

Overventilating. If the indoor and outdoor conditions are very different, such as during winter in northern locations, the amount of ventilation greatly affects the operation of the HVAC system and the energy consumed. Outside air must be heated or cooled before becoming supply air. Therefore, depending on the conditions, ventilating beyond the minimum amount needed for carbon dioxide control may be a waste of energy.

Underventilating. A building space may require cooling even when the outdoor weather is cool. This is because building occupants and electrical equipment, such as lighting, can significantly warm the indoor air. In this situation, the building space can be cooled very efficiently by increasing ventilation with the cool, outside air. *Economizing* is a cooling strategy that adds extra cool outside air to supply air. Not taking advantage of economizing is a missed opportunity to reduce energy consumption in the cooling portion of the HVAC system.

Dirty Filters

Filters are used to trap airborne particulate matter, such as dirt, spores, and pollen, which creates an unhealthy environment for occupants. However, when filters become saturated with contaminants, they excessively impede airflow. This forces the fans in the HVAC system to operate longer and draw more power in order to maintain the desired indoor conditions.

Oversized Fans

In addition to operating at full-speed unnecessarily, many fans are oversized for the required airflow, drawing more power than should be necessary. New airflow analyses and calculations from experienced HVAC technicians are performed to update the HVAC requirements of the facility. These are compared with the actual airflow measurements from the audit to determine if fans are properly sized. If replacing oversized fans is not feasible or cost-effective, the excessive power draw can be reduced by installing variable-frequency drives to slow down the speed of the fan motor.

HVAC SYSTEM AUDIT

The HVAC system portion of an energy audit is likely to be the most involved, due to the size and complexity of the system. A variety of research tasks, inspections, and measurements are typically required.

HVAC System Profile

HVAC systems may include a large number of energy-consuming devices, including boilers, chillers, pumps, fans, and compressors. The first part of an HVAC system audit is to inventory each major load and document its location, operating specification, and description of controls. All of the accessories and controls, such as valves, thermostats, and sensors, associated with each major load should be located and matched up. Much or all of this information should be available in the maintenance management system. This inventory will form the basis for documenting subsequent tests and inspections.

Energy Use Data Logging

For each major load in the HVAC equipment inventory, baseline energy use information should be collected for its normal operation over appropriate business cycles (day, week, or month). The meter must be able to operate for the desired logging period, simultaneously record multiple measurements, and either contain sufficient memory to store the information or connect to another device that can, such as a laptop.

For electrical loads, the measured parameters should include power draw, energy consumption, and power factor. Appropriate logging meters include clamp-on accessories for measuring current and attachable test leads for measuring voltage. Three-phase loads require meters capable of monitoring all three phases simultaneously. **See Figure 3-6.**

Energy Use Data Logging

Fluke Corporation

Figure 3-6. The power draw, energy consumption, and other parameters of each HVAC unit should be logged over time to establish baseline information and determine equipment efficiency.

For other types of energy consumption, meters or gauges can be installed to measure comparable values. For example, a meter in the natural gas fuel system for a boiler measures fuel flow. Logging this information at short intervals shows trends in fuel draw and cumulative consumption.

Since there will likely be more loads to log than available meters, these tasks must be scheduled during the planning phase. Extra time in the schedule may be necessary to be able to retest certain loads if results were corrupted or unusual.

Thermal Imager Scans

All objects emit infrared radiation in proportion to their temperature. Infrared thermometers sense and measure this infrared radiation to determine the temperature of an object. **See Figure 3-7.** Measurements can be taken from a distance without direct contact.

Single-point infrared thermometers display a single temperature reading from a small area. A thermal imager builds and displays an infrared image of an area from many temperature measurements. The electronic image files are transferred to a personal computer for storage, analysis, and inclusion into maintenance reports. Some imagers record both visible light and infrared images of the same field of view, which help identify specific components later. The temperature information is usually shown in false color to highlight the high and low temperatures.

A thermal imager is an extremely effective tool for quickly inspecting a large area or many pieces of equipment and recording the results. Many maintenance problems, such as loose electrical connections, bad motor bearings, moisture penetration, insulation gaps, or leaky ductwork, are immediately apparent with a thermal imager.

Noncontact Infrared-Based Thermometers

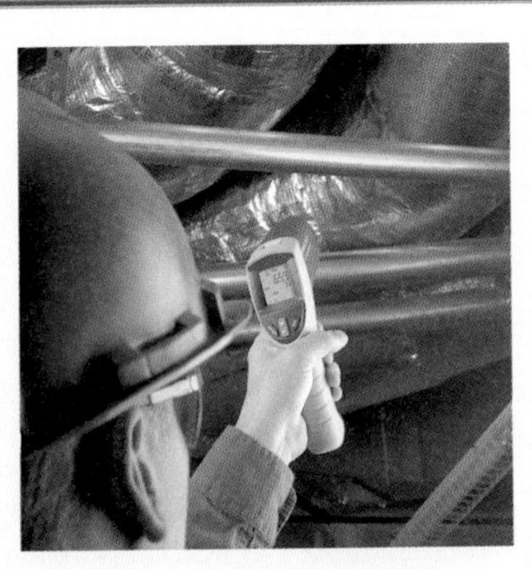

SINGLE-POINT INFRARED THERMOMETER

THERMAL IMAGER

Fluke Corporation

Figure 3-7. Noncontact thermometers measure the infrared radiation of an object, which is proportional to its temperature.

Infrared-based thermometers, of any type, have limitations. Temperature readings may be affected by the surface material and its condition, such as being dirty, painted, or insulated. Shiny materials, such as polished metals, reflect their thermal surroundings, making readings confusing. The surrounding environment, such as water vapor, dust, and gas, may also affect temperature readings.

Building Envelope Inspections. A thermal imager scan is the primary method of identifying leaks in the building envelope. The inspection should cover the entire building envelope, concentrating on doors, windows, expansion joints, and roof surfaces. **See Figure 3-8.** If possible, the inspection should be conducted when the outside temperature is either very cold or very hot. A large difference in temperature between the indoor and outside air makes an air leak stand out prominently on a thermal imager as a sharp color change. Areas of poor insulation typically appear less dramatic, but any significant temperature difference warrants further inspection.

Air Handler and Duct Inspections. Thermal imagers should also be used to inspect any accessible ductwork and air handler equipment for air leaks, which are common at the seams in the sheet metal. Conditioned air from ductwork leaks may remain inside the building envelope but escape into unoccupied areas, such as above drop ceilings, where it is unutilized. Therefore, these types of leaks are another source of wasted energy.

Air Quality Measurements

Indoor air should be tested to confirm that the HVAC system and its controls are achieving the desired conditions specified by the system setpoints. This primarily concerns temperature and humidity and whether sensors are operating correctly. Sensors involved in testing include both room thermostats and sensors inside air handlers and terminal units. The sensor readings should match the test instrument readings within a fraction of a degree or percent. If a sensor does not accurately measure the space conditions, it may need to be recalibrated, relocated, or replaced.

Building Envelope Inspection

Fluke Corporation

Figure 3-8. Thermal imagers are ideal for quickly and precisely scanning large areas, such as the entire building envelope, for leaks and areas of poor insulation.

A number of types of test instruments can take temperature measurements. However, when humidity and other air-quality-related measurements are to be made, it is common to use a meter specialized for indoor air testing. These meters often include accessories that make HVAC-related measurements easier.

Many air quality meters are also able to test for carbon dioxide and carbon monoxide concentrations, which are harmful to occupants. **See Figure 3-9.** A series of carbon dioxide or temperature measurements can be used to calculate the amount of ventilation (percentage of outside air). Some meters perform the calculations automatically after recording the outside-air, mixed-air, and return-air temperatures. The percentage of outside air applied to the total airflow measurement of the air handler yields the volume of outside air. For example, if 40% of an air handler's 12,500 cfm of airflow is outside air, the outside air volume is 5000 cfm.

Air Quality Measurements

Fluke Corporation

Figure 3-9. Air quality measurements, such as for carbon dioxide concentrations, are taken in a variety of locations.

Airflow and Air Velocity Measurements

Airflow is a measure of the volume of air flowing past a point in an interval of time. *Air velocity* is the speed of airflow. When air is constrained within a duct, the airflow equals the air velocity multiplied by the cross-sectional area of the duct.

Airflow and air velocity measurements are usually taken at several points in air handlers, ductwork, and room registers in order to trace the flow of conditioned air. These measurements are also compared to equipment specifications to ensure that the equipment is performing as expected. They are compared also to building requirements to ensure that the equipment is meeting minimum requirements. Furthermore, losses due to leaks can be quantified by measuring air velocity at the leak and multiplying by the cross-sectional area of the opening.

Air velocity meters typically include a thin accessory probe that connects to the meter body. This allows the probe to be inserted into a duct through a small hole while the auditor holds the meter. **See Figure 3-10.** The hole can later be covered with duct tape. After inputting the size of the duct area, the meter calculates and displays the airflow.

Differential Pressure Measurements

Differential pressure is the difference between the pressures on either side of a barrier. These measurements are used to evaluate the effects of filters, fans, dampers, and heating/cooling coils on airflow. For example, a dirty filter restricts airflow, which causes a higher pressure behind the filter. Measuring the pressure difference between the upstream and downstream sides of a filter is an effective way to estimate its condition. **See Figure 3-11.** The pressure differential of a damper can be used to estimate its position, since the airflow varies with damper position.

Air Velocity Measurements

Fluke Corporation

Figure 3-10. The air velocity inside ductwork can be easily measured by meters with thin probe attachments.

A test instrument capable of differential pressure measurements has two ports for connecting air hoses. Small, flexible hoses are connected to the meter to connect each port to the environment being tested. If the meter is located within one of the pressure spaces, that hose can be left off. When the differential pressure mode is activated, the meter displays the measured value. Depending on which environment was used as the reference, the meter may display a positive or negative value. This, and the way the meter is connected, indicates which side is at a higher pressure.

TECH-TIP

Differential pressure is also measured to check whether indoor air is at a higher pressure than outside air. This condition prevents infiltration of unconditioned outside air.

Measuring Filter Condition

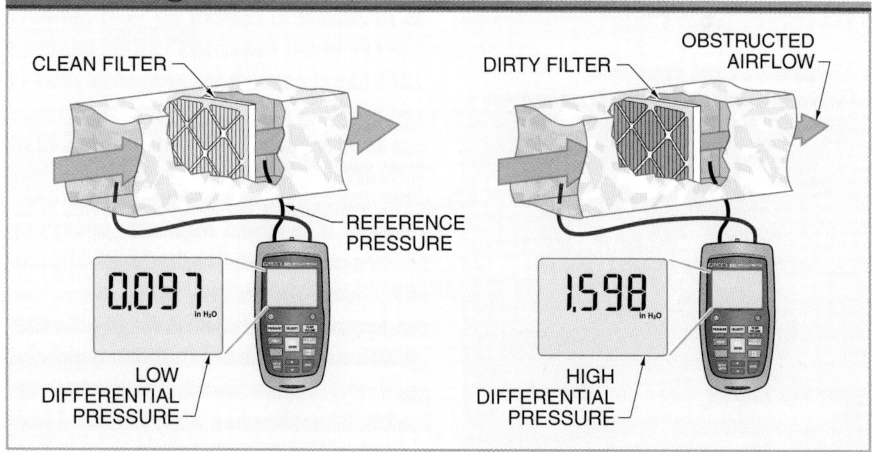

Figure 3-11. Differential pressure measurements are useful for estimating the condition of filters.

chapter 4

Auditing Lighting Systems

Lighting systems consume a significant amount of energy but allow flexible implementation. Changes in lamp type and lighting control can be made relatively easily to reduce energy use without negatively affecting occupant comfort. Some upgrades may improve both energy efficiency and lighting conditions.

LIGHTING SYSTEMS

Proper indoor lighting improves the productivity and safety of building occupants. Lighting should be neither too dark, which can cause eyestrain and safety hazards, nor too bright, which can bother occupants and waste energy. Outdoor lighting at night improves security, illuminates signage, and enhances landscaping features.

Lighting includes both natural daylight and artificial light from lamps. A *lamp* is an electrical output device that converts electrical energy into visible light, along with other forms of energy. Ideally, lighting systems should utilize as much natural daylight as possible and supplement this with artificial lighting as needed to maintain optimal lighting levels for each building space. Artificial lighting is typically a major component of energy consumption for a facility. Auditing lighting systems involves ensuring that lighting is adequate in all areas while identifying ways to minimize the use or energy consumption of artificial lighting.

Lamp Types

Most artificial light is produced by lamps. There are many types of lamps, which are differentiated by the method used to produce light. Common types of lamps used in commercial buildings for lighting applications include conventional incandescent, halogen, fluorescent, and high-intensity discharge lamps. **See Figure 4-1.** A less common lamp type for general lighting is based on a light emitting diode (LED), but this type is becoming more available.

Lamps are installed within light fixtures. A *light fixture* is an electrical appliance that holds one or more lamps securely and includes the electric components necessary to connect the lamp(s) to the appropriate power supply.

Fluke Corporation

Lighting in industrial facilities should be adequate for the tasks being performed but not excessive, which wastes energy.

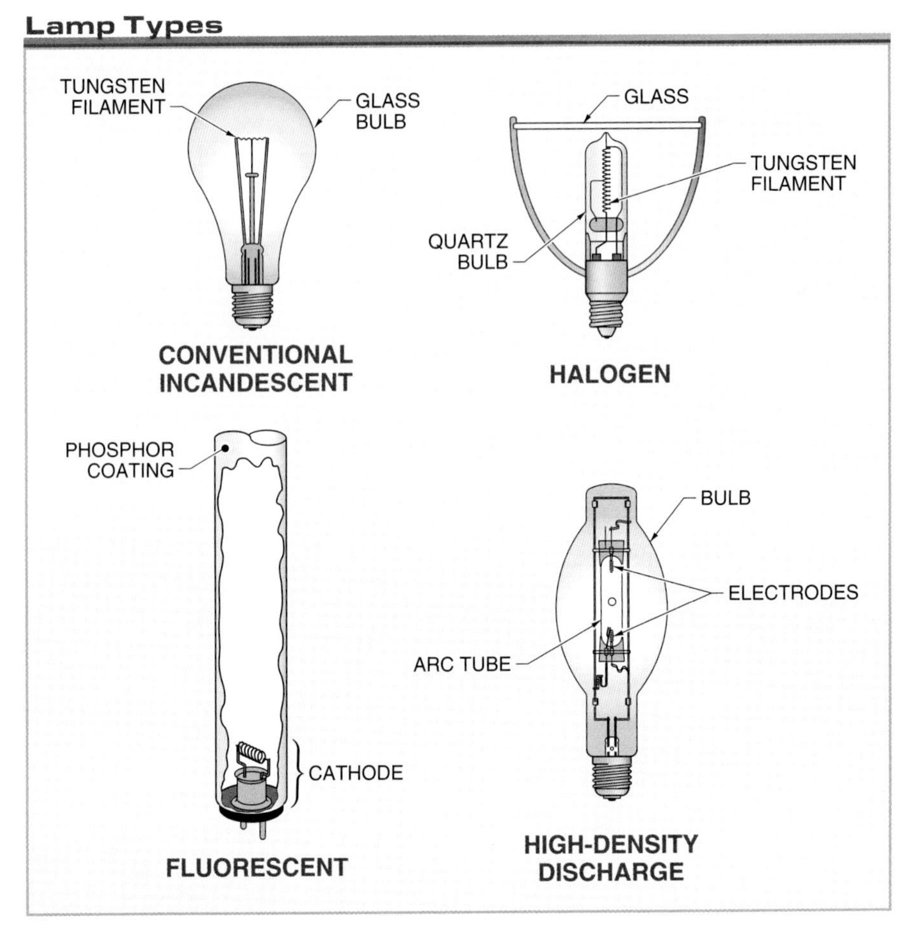

Figure 4-1. There are a variety of lamp types suitable for commercial and industrial facilities. Each type may also include variations based on similar principles.

Incandescent Lamps. Incandescent lamps are the most widely used lamps. An *incandescent lamp* is an electric lamp that produces light by the flow of current through a tungsten wire inside a gas-filled, sealed glass bulb. The glowing filament produces light and a substantial amount of heat as a by-product. Conventional incandescent lamps are relatively inefficient.

A special type of incandescent lamp is a halogen lamp. A *halogen lamp* is an incandescent lamp filled with a halogen gas (iodine or bromine). The halogen gas slows the gradual failure of the tungsten filament, extending the life of the lamp. Because halogen lamps operate at high temperatures, the envelope of the lamp must be made of hard glass or fused quartz. If ordinary glass were used, the high temperatures would soften the glass.

The light output of incandescent lamps is directly determined by the input voltage.

Applying a higher voltage produces more light but severely limits the lifespan of the lamp and is not recommended. However, incandescent lamps are easily dimmed by lowering the voltage.

Gas Discharge Lamps. A *gas discharge lamp* is an electric lamp that produces light by establishing an arc through ionized gas. These lamps must include a ballast in the lighting circuit. **See Figure 4-2.** A *ballast* is a device that controls the flow of current to a gas discharge lamp while providing sufficient starting voltage. Most lamp ballast designs are based on either magnetic inductors or solid-state electronics. Magnetic ballasts are large, heavy, and prone to causing humming noises and lamp flicker. Electronic ballasts produce less flicker and humming effects. Generally, they are also smaller, lighter, and more efficient (and therefore cooler) than magnetic ballasts. They may even be small enough to be built into the lamp. In addition, special electronic ballasts may be capable of dimming lamps.

Ballasts

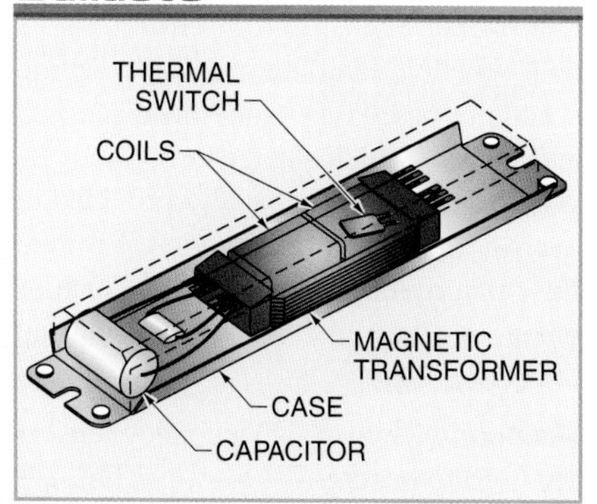

Figure 4-2. Ballasts provide the necessary voltage for starting and operating gas discharge lamps. Some ballasts are also capable of dimming lamps.

TECH-TIP

Since the light output of a lamp changes over time, lamps are often rated in both initial lumens and mean (mid-life) lumens.

A *fluorescent lamp* is a gas discharge lamp that produces light through ionized mercury vapor. The cylindrical glass tube of a fluorescent lamp contains mercury vapor and is sealed at both ends with cathodes. The gas is bombarded by electrons as an arc is formed between the cathodes at each end. This provides ultraviolet light, which causes the phosphor coating on the inner surface of the bulb to fluoresce and therefore, emit visible light. Fluorescent lamps produce very little heat and are relatively cool to the touch.

A *high-intensity discharge (HID) lamp* is a gas discharge lamp that produces light from an electric arc through ionized metal vapor. The arc tube is enclosed within an outer bulb that may include coatings to improve color rendering, increase light output, and reduce surface brightness. There are several types of HID lamps. These types include low-pressure sodium, mercury vapor, metal-halide, and high-pressure sodium lamps. Each type of lamp uses a different metal, which determines the starting, operating, and light output characteristics.

Light Emitting Diodes. A *light emitting diode (LED)* is a semiconductor device that emits a specific color of light when DC voltage is applied in one direction. They have extremely long life spans and are resistant to physical shock and vibration. Although, some LED-based lighting products are available, LEDs are not yet common for general lighting applications. However, further growth is expected.

Individually, LEDs produce little light output, but their power consumption is so low that their efficiency is comparable to halogen and fluorescent lamps. A large number of LEDs are grouped together into an array to provide the necessary lumen output for general lighting. Most LED lighting devices require an LED driver, which functions similar to a ballast.

Lighting Control

Switching is the application of electrical power to a lamp and dimming is the modulation of the amount of power applied to a lamp, changing the light output. These modes can be used separately or together in lighting control systems. Lighting circuits can be controlled manually, but this is usually not the most efficient method of optimizing energy use. Occupants can easily leave artificial lighting on when it is not needed, such as when natural light is adequate or the area will be unoccupied. Therefore, many commercial and industrial facilities use automatic switching and dimming control systems to save energy. There are several different types of control systems that can control lighting circuits, but most provide similar control features.

Schedules. The simplest and often most cost-effective method of lighting control is based on time schedules. A central lighting controller is programmed with a series of times of day that represent when lights should be switched ON and OFF. Schedules are used to restrict lighting to the times of day that occupants are expected to be present, such as during normal working hours. Schedules are also useful for automatically operating outdoor lighting overnight. **See Figure 4-3.**

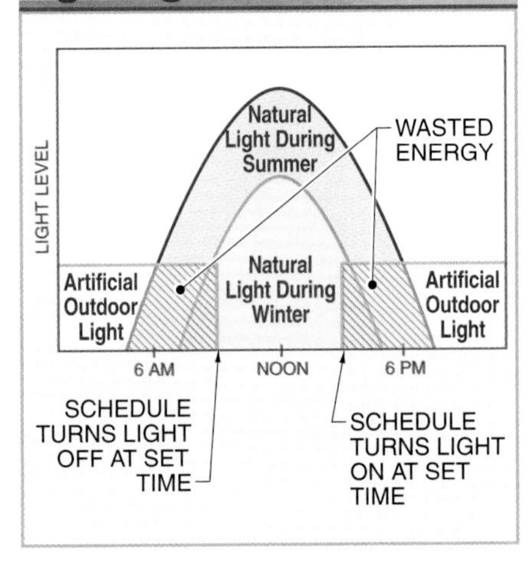

Figure 4-3. Controlling a lighting circuit with fixed schedules does not allow for seasonal changes in natural lighting.

Some schedule-based controllers are relatively simple in that they use only one set of fixed times to control all lighting simultaneously. Others provide sophisticated capabilities for setting different schedules for weekdays, weekends, and holidays; controlling each lighting circuit by its own schedule; and automatically setting schedules based on astronomical sunrises and sunsets, which change throughout the year. Some means is usually provided to override a schedule-based control system for extraordinary circumstances, such as when an area must be accessed during off-hours.

Occupancy Sensors. Occupancy sensors are used to ensure that artificial lighting is used only when needed. In most circumstances, when there are no occupants in an area, the lighting can be switched OFF. An *occupancy sensor* is a sensor that detects the presence of at least one person in an area. There are a few different types of sensors, which detect people by their body heat, motion, or both. The sensors are typically mounted high on walls or ceilings in order to sense a wide area.

Occupancy sensors are used to switch one or more lighting circuits ON when an occupant is detected. As occupants remain in the area, the sensor continues to detect their presence and keeps the lighting ON. When all occupants leave the area, the sensor no longer detects occupancy. Usually, the controller is programmed to delay switching the lights OFF for a certain interval, which minimizes frequent switching for areas where people move into and out of repeatedly.

The use of occupancy sensors usually results in less operating time for lamps than scheduling alone. It ensures that lighting is activated only when needed. However, this type of control is not appropriate for all applications. Some lamps require several seconds to a few minutes to fully illuminate, so it is recommended that they be left ON for longer periods of time. Also, occupancy sensors may not be practical for very large areas, such as warehouses or manufacturing buildings.

Ambient Light Level. Measuring the ambient light level determines whether artificial lighting is needed at all for an area. If sufficient light is available through natural lighting, then the lighting circuits will not be turned ON. As the sun sets in the evening, the measured light level falls, and when it reaches a certain threshold, the lighting controller energizes the circuits. Artificial lights may operate until the morning daylight raises the natural light level above the threshold and the circuits are turned OFF.

Simple ambient light level sensors are often used to control outdoor lighting. A similar ON/OFF timing of the lighting circuits can be achieved with schedule-based controls but with two disadvantages. Simple schedules may not be able to adjust to seasonal changes in sunset and sunrise times. Also, systems with light level sensors can activate lighting as needed during the day, such as during very overcast or stormy weather.

Daylighting. A light-level-based control strategy can be used for indoor lighting, but it requires more sophisticated control than simple ON/OFF switching. Natural daylight is exploited to minimize the use of artificial lights. Lights are turned OFF when not needed or dimmed at various levels throughout the day to provide adequate additional light without excess.

Indoor applications call for a smooth transition from natural to artificial light. Light level sensors measure the total amount of illumination in a space (from all sources). The system adds only the amount of artificial light needed to maintain the desired illumination. **See Figure 4-4.** Daylighting systems also incorporate schedules and/or occupancy sensors so that the daylighting function operates only when there are occupants in the building.

TECH-TIP

Lighting controls ensure that spaces are lit adequately, which improves occupant productivity and the safety and security of the occupants' surroundings. These systems operate automatically in the background without operator input, adding convenience and saving time.

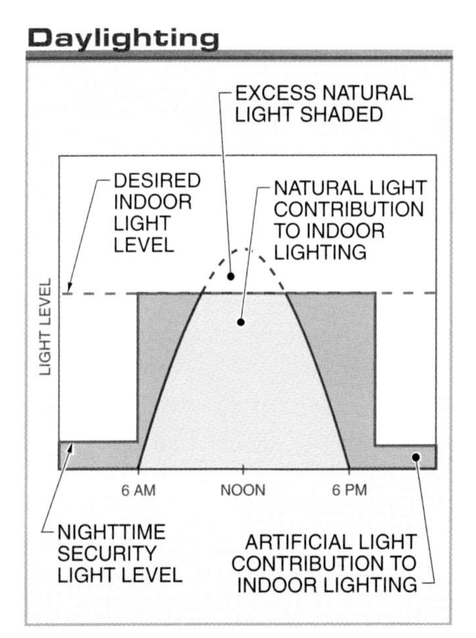

Figure 4-4. Daylighting adds only enough artificial lighting to maintain a desired light level.

LIGHTING SYSTEM INEFFICIENCIES

Lighting systems are a prime target of energy audits because they are typically a major consumer of electricity and there are several potential sources of inefficiency. These opportunities for energy savings involve the choice of lamp types and their use along with determining how lighting can negatively impact the energy use in other building systems.

Lamp Luminous Efficacy

Only a relatively small percentage of the total energy produced by a lamp is visible light. Approximately 65% to 90% of the total energy produced by lamps is ultraviolet energy, infrared energy, or heat, which are not visible. For example, only about 21% of the total energy output of a typical fluorescent lamp is visible light.

Visible light is measured in lumens. A *lumen (lm)* is a unit used to measure the intensity of visible light from a light source. Lamps are rated in watts and lumens, which provide a measure of efficiency. *Luminous efficacy* is a ratio of the light output of a lamp to the electrical power input. This information is used to compare lamps and determine the power consumption and cost of certain lighting levels. For a general lamp type, the luminous efficacy is given as a range because larger wattage lamps within the same type tend to be more efficient. **See Figure 4-5.** For providing a certain level of illumination, lamps with a higher luminous efficacy are less expensive to operate.

Lamp Electrical Efficiency

Lamp Type	Luminous Efficacy*
Conventional incandescent	12 to 18
Halogen	18 to 24
Fluorescent	55 to 100
Low-pressure sodium HID	190 to 200
Mercury vapor HID	50 to 60
Metal-halide HID	65 to 115
High-pressure sodium HID	100 to 150
Light emitting diode	25 to 70

* in lm/W

Figure 4-5. Luminous efficacy is a measure of how efficiently a lamp produces light for each unit of electrical power.

Illumination Standards

Light illuminates surfaces. *Illuminance* is the quantity of light per unit of surface area. **See Figure 4-6.** Illuminance is measured in foot-candles (U.S. customary system) or lux (metric system). A *foot-candle (fc)* is the illuminance from 1 lumen per square foot (1 lm/ft^2) of surface. A *lux (lx)* is the illuminance from 1 lumen per square meter (1 lm/m^2) of surface. For example, a lamp in the center of a room illuminates the walls, floor, and ceiling. The average illuminance in the room is the total light output from the lamp divided by the total area of all surfaces illuminated.

Illuminance

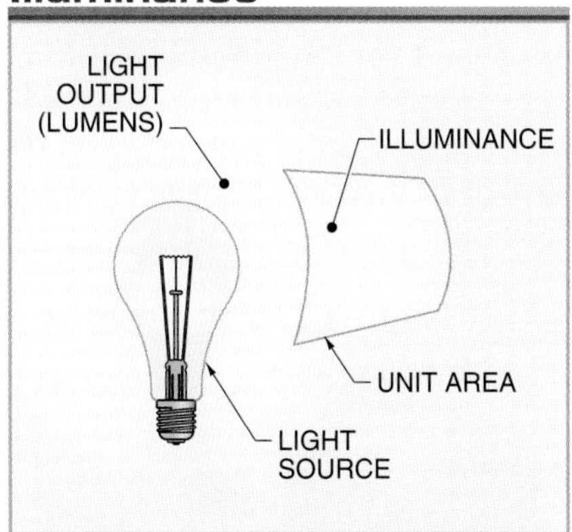

Figure 4-6. The total light output of a lamp is measured in lumens while illuminance is the density of that light on a surface.

Some building codes and other regulations include minimum illuminance levels for workplaces and venues. Depending on the jurisdiction, these specifications may be requirements or recommendations. The illuminance levels depend on the type of area and the tasks being done in that area. For example, general office work in an office building requires about 100 fc to 200 fc of illumination. Storage areas in a warehouse may need only 30 fc, but detailed assembly areas in manufacturing buildings may require up to 300 fc.

An energy audit often includes measuring the illuminance levels in various areas of the facility. This data is compared with the minimum illuminance standards for the task being done in each area to determine if the lighting system should be changed. While it is possible that additional lighting is needed in some areas, which may increase energy consumption, it may also be determined that other areas are already lit more than necessary. Adjusting the illumination in these areas to provide adequate, but not excessive, lighting may result in a net reduction in energy consumption.

Heat Gain

Lamps add heat to a building space as a waste product. While this can be beneficial during the heating months, it is a relatively inefficient way to generate heat and is counterproductive during the cooling months. This heat gain must be considered a thermal load that could affect the operation and energy usage of the HVAC system.

Also, using a lot of natural light indoors requires a significant window area, which affects the thermal balance of the building. Direct sunlight can heat a space, which adds an extra cooling load in the summer. Window glass also allows some indoor heat to escape during the winter and therefore, more heat is required to warm the building. New construction designed for daylighting may include architectural features or special window treatments to manage sunlight infiltration. Daylighting systems may also involve controlling motorized window blinds or curtains to balance the impacts of lighting and HVAC systems on energy consumption.

Daylighting measures may include the use of permanent window shades that shield workstations from direct sunlight but reflect some of the natural light onto the ceiling.

Power Quality

Certain lighting components, such as ballasts or electronic controls, may affect the power quality of an electrical distribution system. For example, inductive elements in ballasts lower the electrical supply power factor. This reduces the amount of useful electrical power available and requires increased conductor and equipment sizes. Poor power quality may also damage equipment and the electrical distribution system itself. When proposing changes to the lighting system, the effect of those changes on the power quality of the facility should be taken into consideration.

LIGHTING SYSTEM AUDITS

An investigation of the lighting system provides more than just energy savings information. By measuring light levels in various areas, the audit can determine whether lighting is adequate for the tasks being performed. Improving the lighting system can boost productivity and identify issues that may affect other systems, such as the HVAC system.

Illuminance Level Measurements

Illuminance is easily measured with handheld light meters. These test instruments consist of a sensor installed beneath a small, white dome and the electronics needed to process and display readings. **See Figure 4-7.** Taking a measurement involves placing the sensor near a work area, such as a desk, and pointing it toward the view of the occupants (up or to the side).

Measuring Light Levels

Fluke Corporation

Figure 4-7. Light levels are easily measured with portable light-level test instruments.

Distance from the light sources affects the reading, so the sensor should be placed where tasks are completed. Since the sensor should not be shadowed by the auditor, many models allow the sensor portion to be detached and held at a distance from the body. Meters may include special functions such as holding a reading on the display or recording the minimum and maximum values of changing illuminance.

Lamp Type Identification

Many types of lamps are suitable for commercial and industrial applications. When auditing a lighting system, the type, size, and number of lamps in each fixture should be noted as well as the location and type of each fixture. Information about each lighting area should also include special considerations, such as minimum lumen requirements for the tasks and whether true color rendition is important. Lighting component manufacturers can provide specifications for light output, luminous efficacy, and other characteristics.

By analyzing the requirements and allowances of each area or application, it may be determined that energy use can be reduced by using a different lamp type and/or relocating the fixtures. For example, if an outdoor parking area uses fluorescent lamps, switching to low-pressure sodium HID lamps would reduce energy use. These lamps produce a yellow light that is objectionable for most indoor applications but acceptable for many outdoor applications. In this situation, the increased efficiency is more important than the light color.

Light Fixture Temperature Measurements

A certain amount of heat by-product from lamp operation is normal, but excessive heat indicates a problem with lighting components or electrical connections. Ballasts for gas discharge lamps are particularly prone to overheating. As each light fixture is inspected for lamp information, a thermal imager or noncontact thermometer should be used to measure temperatures. Ballasts, junction boxes, controllers, lamp connectors, and other lighting components should be checked and the temperature information carefully documented. Normal operating temperatures can be found in literature from the manufacturer for comparison.

Control Strategy Identification

Changing a lighting system control strategy, such as from manual control to occupancy-sensor-based control, can be a significant investment but can also result in significant energy savings. The costs and expected savings of various control options are calculated later from the information gathered during the audit. However, the auditor should consider these possibilities while reviewing the lighting needs and conditions for each area. For example, the auditor may discover that based on the way a particular group of rooms are used, their lighting is well-suited for upgrading to occupancy-based control. This type of information should be noted on auditing forms as appropriate.

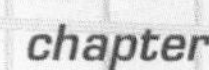

chapter 5

Auditing Electrical Systems

The two most significant building systems, HVAC and lighting, are major parts of the electrical system. However, the energy-consuming equipment in these systems is often dealt with separately in an energy audit. The electrical system audit concentrates on the remainder of the large electrical loads.

ELECTRICAL SYSTEMS

An *electrical system* is a combination of electrical devices and components, connected by conductors, that distributes and controls the flow of electricity from its source to a point of use. At that point, electricity is converted into some type of useful output, such as motion, light, heat, or sound.

Electricity Distribution

An *electrical service* is the AC electrical power supply to a building or structure. Power is supplied to industrial and commercial facilities at high voltages that must be reduced before use. A distribution system within a building then delivers power from the electrical service to end-use points throughout the building. From these points, electricity powers individual loads, such as motors, lamps, and computers. **See Figure 5-1.**

A *switchboard* is a set of equipment that breaks down the incoming electrical power of a utility into smaller units for distribution throughout a facility. The switchboard is the last point on the power distribution system for the power company and the beginning of the power distribution system for the electrician of the property. The switchboard contains overcurrent protection devices and switches to control the flow of electricity into the building. A switchboard may also contain equipment for controlling, monitoring, and protecting the electrical use in a building.

From the main switchboard, busways are used to distribute electricity for feeder circuits. A *busway* is an electricity distribution system composed of enclosed busbars. Individual circuits are taken off the busway through panelboards. A *panelboard* is a wall-mounted distribution cabinet containing a group of overcurrent protection devices and switches for branch circuits. A *branch circuit* is the circuit in a power distribution system between the final overcurrent protection device and the associated end-use points, such as receptacles or loads.

A *transformer* is an electrical device that increases or decreases AC voltage, with a proportional and opposite effect on current. Transformers are used at various points within the power distribution system of a building to lower the voltage to the levels required by most loads, such as 480 V, 240 V, 208 V, 120 V, or 24 V.

TECH-TIP

The U.S. industrial sector consumes more than 700 billion kWh of electricity, and spends more than $30 billion annually, for motor-driven systems.

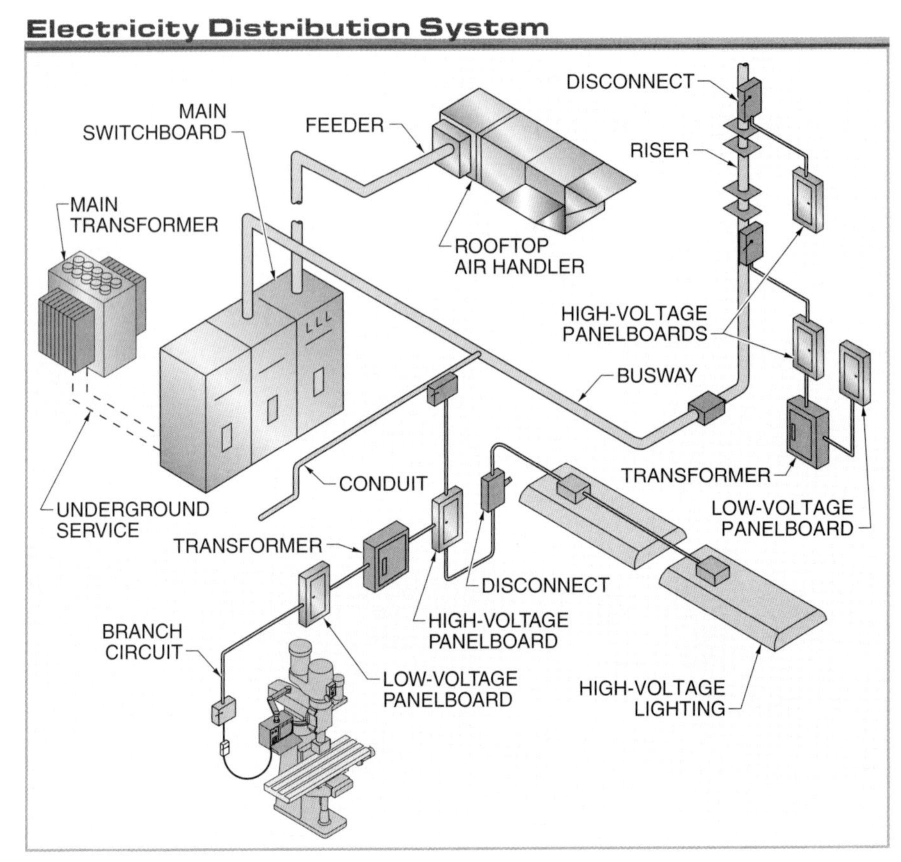

Figure 5-1. Electricity is distributed through a facility using a system of transformers, switchboards, busways, panelboards, and other devices.

Motors

Industrial facilities use many AC motors throughout the facility. Motors can be designed to operate on either single-phase (1ϕ) or three-phase (3ϕ) power. Single-phase motors are used for light loads and may not even be represented in an energy audit. Three-phase motors are far more common in industrial facilities and are usually so large that they can have a significant effect on facility efficiencies.

A motor rotates due to the interaction between the magnetic fields of the stator and rotor. A *stator* is the stationary frame that surrounds the rotor of an AC motor. A *rotor* is the rotating part of an AC motor. The rotor is mounted on the motor shaft, which is attached to the load of the motor. The direction, speed, torque (rotational force), and other operating parameters of a motor are determined by motor controls and/or variable-frequency drives.

Motor Controls. Motor controls are the collection of sensors, switches, timers, relays, and other devices for automatically operating a motor according to a set of criteria. For example, a motor that runs a pump must start when a pressure falls to a minimum setpoint and stop when the pressure rises to a maximum setpoint. A motor control circuit using pressure switches and relays can control this action automatically. **See Figure 5-2.**

Motor Controls

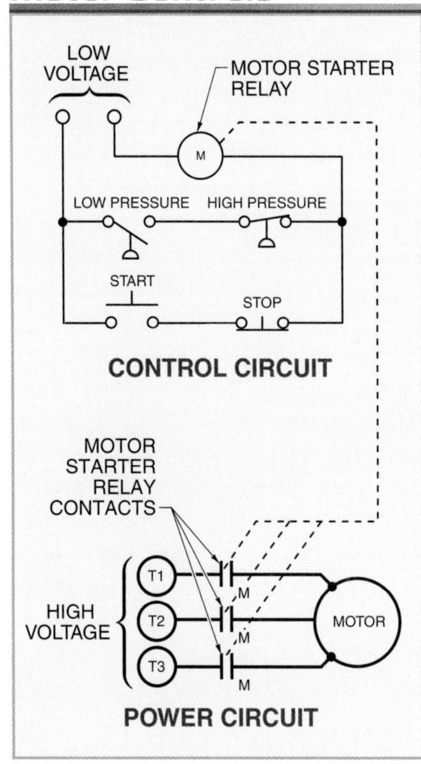

Figure 5-2. Motor controls include sensors, switches, relays, and other devices connected in such a way as to automatically run and stop a motor according to circuit logic.

A wide variety of devices can be used as part of a motor control circuit. The way in which the devices are connected also provides circuit logic, such as requiring that multiple criteria be met before the motor runs.

Variable-Frequency Drives. The frequency of the AC power supply determines the speed of a motor. Since normal power line frequency is constant, it must be changed to increase or decrease the motor speed. A *variable-frequency drive* is a motor controller that is used to change the speed of an AC motor by changing the frequency of the supply voltage. The drive uses very high frequency pulses of varying length to simulate a lower-frequency sine wave, which is the desired operating frequency. **See Figure 5-3.** By changing the length and polarity of the pulses, sine waves of almost any frequency can be simulated.

Variable-frequency drives allow a motor to be operated at almost any speed. The speed can also be changed gradually, allowing the drive to control motor acceleration and deceleration. Drives may also feature many motor control functions, motor torque control, and motor braking.

ELECTRICITY LOSSES

Electricity is the most significant type of energy used by a building or facility. A very large number of devices use electricity and draw power in a variety of ways. Equipment that is older, poorly maintained, improperly sized, or not optimally controlled tends to use too much electricity or use it inefficiently. Some equipment can even adversely affect the efficiency of other devices by causing poor power quality.

High Electricity Use

Electricity is billed according to the total consumed, so reducing this consumption is a major part of reducing utility costs. However, for industrial customers, a significant portion of an electricity bill is usually related to the rate and time of day at which the electricity is consumed. Changes in the power demand can also reduce costs.

Electricity Consumption. *Electricity consumption* is the total amount of electricity used during a period of time. Electricity consumption is measured in kilowatt-hours (kWh). Electricity consumption is metered for determining the amount of electricity delivered to (or from) a facility for billing purposes. Meters are installed at the facility service entrance and establish the transition between utility and customer-owned equipment.

Power Demand. Utility customers commonly use more electricity during some periods of the day than others. *Power demand* is the amount of electrical power drawn by loads at a specific moment. Power demand is constantly changing and is measured in kilowatts (kW).

High power demand makes it more difficult for electric utilities to supply power to all of their customers. Electricity during high-demand periods is more valuable, so the utility often charges extra. An electric bill for an industrial facility may reflect the changing rates for electricity, sometimes on an hourly basis. Alternatively, the utility may include extra charges for high demand or base a bill for an entire month on the rate for the highest demand during that period. In any of these billing scenarios, either reducing the overall energy use or shifting loads to off-peak times can significantly reduce utility costs.

Poor Power Quality

Power quality is a measure of how closely the power in an electrical system matches nominal (ideal) characteristics. It is common for actual electrical parameters to vary somewhat, but allowable ranges are typically very small. Good power quality means that the parameters are within acceptable limits for the electrical system. Poor power quality has excessive variations in the parameters, which can cause damage to loads and circuit equipment.

Power quality problems can be caused by the electrical generation and distribution equipment, but they can also be caused by loads operating on the electrical system. Power quality can involve many characteristics of the electrical supply, such as voltage, harmonics, power factor, and unbalanced conditions in three-phase power supplies. These can all be monitored and conditioned if improvement is necessary.

Variable-Frequency Drive Output

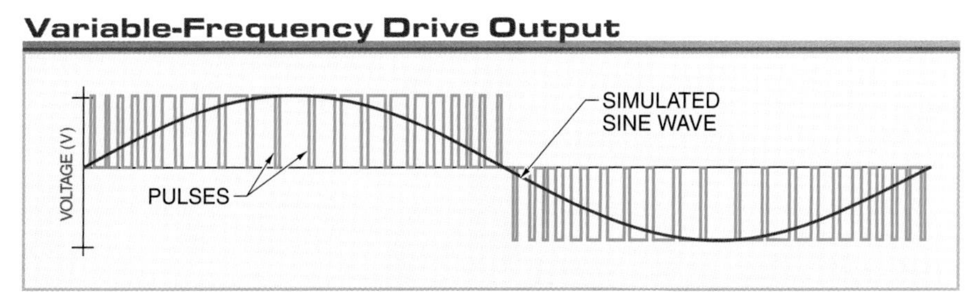

Figure 5-3. Variable-frequency drives supply power to motors in the form of a very high frequency series of pulses. By adjusting the length and polarity of the pulses, drives can control all aspects of motor operation.

Voltage Variations. Voltage in a power distribution system is typically acceptable within the range of +5% to –10% from the nominal voltage. Small voltage fluctuations typically do not affect equipment performance, but voltage fluctuations outside the normal range can cause circuit and load problems. **See Figure 5-4.** Computer equipment is particularly sensitive to noise or voltage fluctuations.

Voltage sags (reductions) are commonly caused by overloaded transformers, undersized conductors, conductor runs that are too long, too many loads on a circuit, peak power usage periods (brownouts), and high-current loads being turned ON. Voltage sags are often followed by voltage swells as voltage regulators overcompensate.

Voltage swells (increases) are often caused by large loads being turned OFF. Lasting overvoltages are sometimes caused by incorrectly wired transformers or being near the beginning of a power distribution system. Voltage swells are not as common as voltage sags but are more damaging to electrical equipment.

A *transient* is a very brief and very large voltage spike. Transient voltages are caused by lightning strikes, unfiltered electrical equipment, contact bounce, arcing, and high-current loads being switched ON and OFF. Transients differ from voltage swells by being larger in amplitude, shorter in duration, steeper in rise time, and erratic. High-voltage transients can permanently damage circuits or electrical equipment.

Harmonics. A *harmonic* is a waveform component at an integer multiple of the fundamental waveform frequency. For example, the second harmonic frequency of a 60 Hz sine wave is 120 Hz, the third is 180 Hz, and so on. These higher-frequency harmonic components superimpose on the fundamental frequency, distorting the waveform. **See Figure 5-5.**

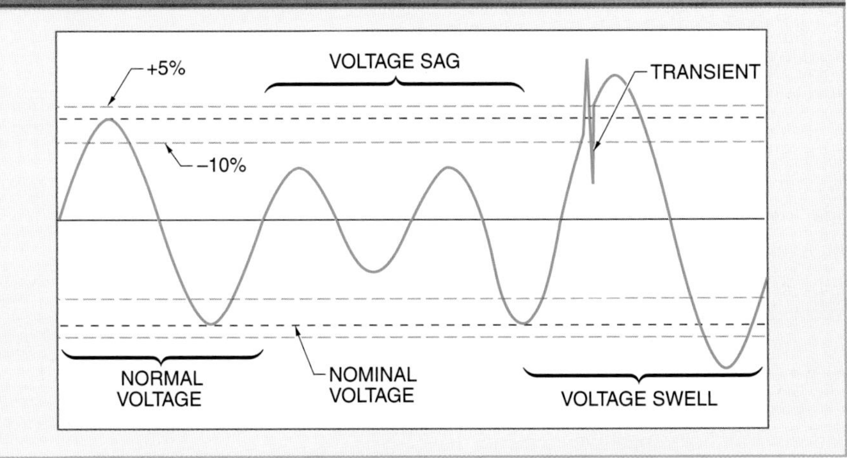

Figure 5-4. Power supply voltages normally remain within the acceptable range of +5% to –10% of their nominal value. Voltage variations outside of this range can cause power quality problems.

Harmonics

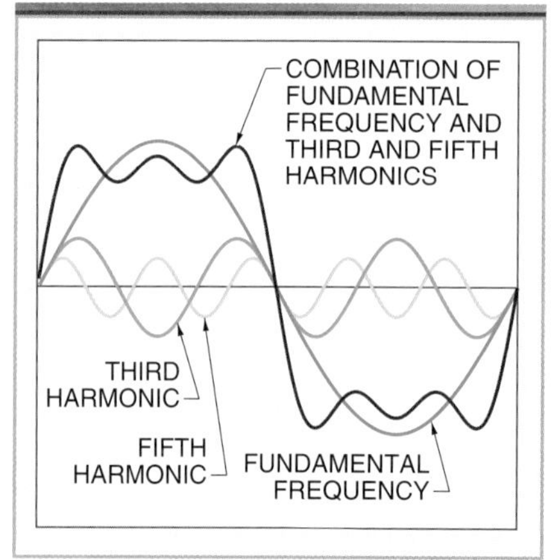

Figure 5-5. When harmonics combine with the fundamental frequency, the resulting distorted waveform is a power quality problem.

Harmonics are commonly caused by nonlinear loads, such as variable-frequency drives and switching power supplies. Harmonics cause extra heat in motors and transformers and sometimes create audible noise.

Low Power Factors. AC loads are either resistive or reactive loads. A *resistive load* is a load that keeps voltage and current waveforms in phase. *True power* is the product of in-phase voltage and current waveforms that produces useful work. **See Figure 5-6.** True power is represented in units of watts (W) and is converted into work, such as sound produced by speakers, rotary motion by motors, light by lamps, linear motion by solenoids, and heat by heating elements.

A *reactive load* is an AC load with inductive and/or capacitive elements that cause current and voltage waveforms to become out of phase. Inductive loads are the most common reactive loads and include motors and transformers. *Reactive power* is the product of out-of-phase voltage and current waveforms that does not result in a net power flow. Inductive loads retard current in the process of building magnetic fields, causing the current waveform to lag the voltage waveform. Capacitive loads store voltage, causing the current waveform to lead the voltage waveform. Reactive power is represented in units of volt-amperes reactive (VAR).

Power Factor

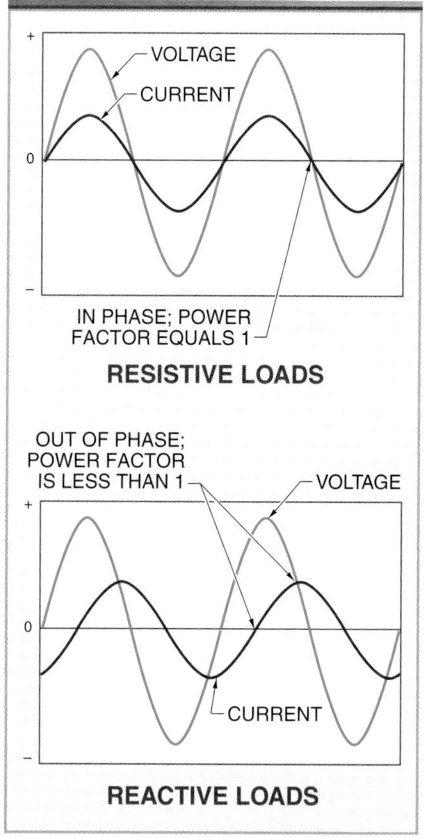

Figure 5-6. The power factor is a measure of how far current and voltage waveforms are out of phase.

Power factor is the ratio of true power to apparent power, which describes the displacement of voltage and current waveforms in AC circuits. *Apparent power* is a combination of true and reactive power and is given in units of volt-amperes (VA). For resistive loads, voltage and current waveforms are in phase and apparent power equals the true power, so the power factor equals 1. Reactive load circuits have a power factor of less than 1 because the true power is less than the apparent power.

When the power factor is less than 1, the circuit is less efficient and has a higher operating cost because not all current is performing work. Because reactive loads return some power to the source, larger conductors, overcurrent protection, switchgear, and other distribution equipment must be provided for loads with lower power factors. Consequently, maintaining a high power factor minimizes the sizes of and costs for the equipment.

Load Unbalance. A three-phase electrical system may be used also to power single-phase loads, but they must be balanced across all three legs. Unequal loading of the conductors and overcurrent protection devices can cause overheating and damage. Unbalanced phases can also damage the three-phase loads on the system.

Unbalance may occur in the voltage, current, and/or phase spacing of the system. Voltage unbalance is the maximum voltage difference between any two phases and should not be more than 1%. Current unbalances should never exceed 10%.

Motor Inefficiency

Electrical system audits determine the efficiencies of major electrical loads, which include many motors. Motors become less efficient and fail, sometimes gradually and sometimes rapidly, due to a variety of causes. The most common causes are overloading and contamination with dust, water, oil, or grease. Other causes include poor power quality, misalignment, lack of lubrication, mechanical failures, frequent stopping and starting, and gradual wear. Many of the causes are related to overheating, which melts the insulation on the stator coil wires, causing shorts and damaging the motor.

The oldest motors tend to be the least efficient. If a motor is unacceptably inefficient, or trending that way, auditors must consider possible remedies for improving its performance. Options for improving performance include repair, replacement, and upgrades with motor drives or other controls. However, some motor problems, particularly power quality issues, can be solved by addressing problems in other equipment and may not require changes to the motor.

It is common for motors to be improperly sized for their intended use. Undersized motors are overloaded and eventually fail due to overheating damage, which causes excessive maintenance and replacement costs. Most motors can handle brief overloads without damage, but sustained current draw in excess of 105% of the nameplate rating indicates an undersized motor. Oversized motors draw too much power for the work required, which is energy waste. Instead, drives can be installed that operate an oversized motor at a lower speed to accomplish the same work while drawing less power.

Excessive Heat

Electrical connections should be clean and tight, which minimizes resistance to current flow. Connections that are loose, overtightened, or corroded have higher resistances, causing a voltage drop across the connection that manifests as heat.

The heat itself is a waste of electricity, but perhaps more importantly, it may be an indication of more serious electrical problems. For example, voltage unbalance between phases in a three-phase system can cause excessive heat in panels and other enclosures. Very hot fuses indicate that the current flow is near the fuse rating and the circuit is overloaded.

However, some amount of heat is normal for many connections, as resistance cannot be completely eliminated. Therefore, a temperature below that of comparable connections may also indicate a problem, such as an open in a circuit that is preventing any current flow.

ELECTRICAL SYSTEM AUDITS

Electrical work and measurements involve special considerations. If power must be shut OFF to connect meters, the building systems and industrial processes may be affected. These tasks must be carefully planned and approved by management.

Personnel connecting the meters or taking electrical measurements must be qualified for this type of electrical work. Appropriate personal protective equipment (PPE), which varies according to the voltage level, task, and distance to live parts, must be used. **See Figure 5-7.** If qualified personnel and necessary equipment are not available, outside consultants may be hired to perform the audit.

Extra safety precautions are needed if electrical panels cannot be normally closed and secured during logging. For example, a boundary must be established at the necessary distance from live parts so that personnel cannot enter the restricted space without the proper PPE.

Electrical PPE

Fluke Corporation

Figure 5-7. All electrical work, including energy audit measurements, must be completed by qualified personnel using the required personal protective equipment (PPE).

Electrical Load Inventory

The first step in an electrical system audit is to inventory the major electrical loads not already audited with the HVAC and lighting systems. Equipment specifications and documentation on the expected energy use of each load must be collected. Also, equipment maintenance records must be reviewed for the preceding 12 months. Loads with higher than average incidences of failure or other problems must be identified. These will likely be sources of waste and possible savings opportunities.

Energy Use Data Logging

The baseline energy requirements must be established for each major electrical load in the inventory. The electrical supply to each load is logged to record power demand, cumulative electrical consumption, and power factor over a complete business cycle. Other important parameters to measure may include harmonics, voltage fluctuations, transients, and three-phase unbalance. Depending on the facility, a business cycle could be a day, week, or month. When planning an audit, sufficient time must be allowed to log each load for a full cycle.

Also, these same electrical parameters should be logged at each electrical service entrance and major distribution panel of the facility. The utility meter reading should be noted at the beginning and end of these logs. The logs provide electrical use data for entire buildings or systems and help establish an overview of the major streams of electricity use. **See Figure 5-8.**

Some meters are dedicated to extensive data logging applications, while some multipurpose, handheld meters have limited data logging functions. **See Figure 5-9.** The choice of meter for each application depends primarily on the measurement capabilities and memory capacity of the meter.

The meter is connected at the appropriate panel to record energy consumption for the loads downstream of that point. Voltage probes include clips so that they can be secured to the test point. Current clamp accessories are connected for measuring current information. (Depending on the meter, voltage and current may or may not be simultaneously recordable.) Data logging parameters, such as the measurements to record, logging interval, and total log time, are programmed into the meter. Measurements and their associated timestamp are recorded onto on-board memory. They can be displayed on the screen as graphs and downloaded later onto computers for analysis.

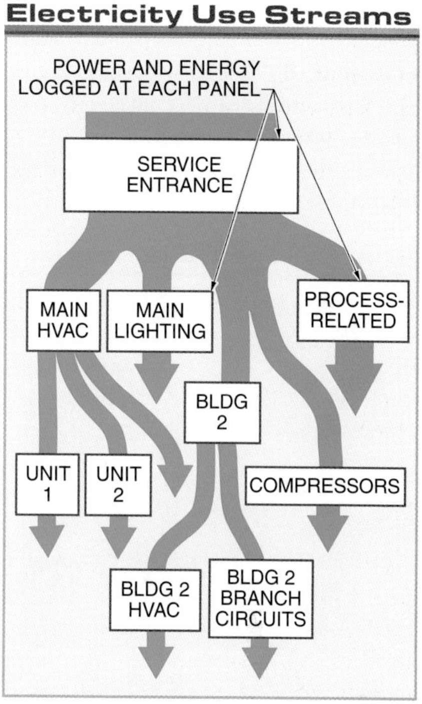

Electricity Use Streams

Figure 5-8. Power and energy logging at the service entrance and various other panels reveals how electricity is used in various parts of a facility.

The meter's internal batteries can operate it for short periods of time. If logging for more than a few hours, it will probably be necessary to connect the meter to an external power supply. The meter documentation provides guidance on power supply options.

Three-phase circuits should be logged with the appropriate meter that can connect voltage probes and current clamps to all three phases (and also the neutral conductor, if needed) simultaneously. Auditors must be careful to place the probes on the correct phase, as indicated by the labels. **See Figure 5-10.** These meters record comprehensive power and energy parameters for the entire three-phase system, such as load unbalances that affect power quality. However, some limited tests on three-phase systems can still be performed by simpler, single-phase meters, such as identifying voltage fluctuations that are interfering with equipment operation.

Energy Use Data Loggers

Fluke Corporation

Figure 5-9. Data logging meters must be able to measure the desired electrical parameters at intervals over a period of time and store the recordings for later downloading and analysis.

Electricity Billing Review. The utility meter and utility bill should be checked for accuracy against the data from the electrical service entrance. If the bill lists a peak demand charge or penalty, it should be determined how this was calculated, and the logged data should be used to identify the moment of peak demand. The cause of the demand must be determined, and ways to reduce demand at that time of day should be investigated. Similarly, if the bill lists a power factor penalty, the power factor level that triggered utility charges must be determined. The data logs must be reviewed, and the loads with a low power factor must be identified.

Efficiency Evaluations. Baseline data is used to evaluate the efficiency and operating trends of the major electrical loads. The original electrical requirements (from equipment specifications) must be compared to current electricity use to determine if a load has become less efficient over time.

The loads that are no longer operating within their tolerance specifications must be flagged. Information should be collected on possible replacements. Efficiency data may also identify opportunities to add variable-frequency drives or other controls to adjust motor output to match demand, thereby reducing energy consumption. If motor controls are already in place, they should be evaluated for further improving their logic or operation for reducing motor energy use.

TECH-TIP

When measuring current, power can be estimated by multiplying current by voltage. However, this assumes a perfect power factor, so this estimate should be used only for approximations.

Three-Phase Data Logging

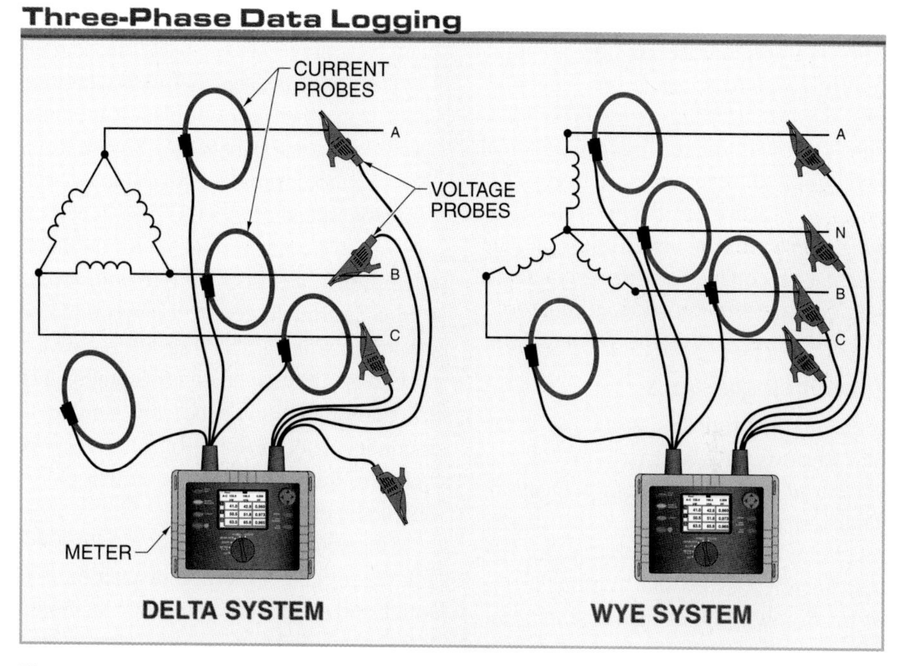

Figure 5-10. Logging three-phase power systems requires the use of meters that can connect to all three phases of the systems simultaneously.

Current Measurements

Additional measurements of electrical current may be needed to complete the profile of significant electricity use throughout a facility. Also, current measurements may be needed to identify sources of unusual current draw discovered during power data logging. These measurements systematically test and eliminate circuits until the problem is identified.

Using a clamp-on ammeter is a convenient and safe way to measure current flow without opening a circuit or touching any live parts. A clamp-on ammeter measures current by sensing the resulting magnetic field. **See Figure 5-11.** To take a current measurement using a clamp-on ammeter, the meter is first turned ON. Next, the appropriate type of current being measured is selected, AC or DC. Finally, the jaws of the clamp are opened and closed around the conductor.

Clamp-On Ammeters

Fluke Corporation

Figure 5-11. Using a clamp-on ammeter is a safe and convenient way to measure current through a conductor.

The jaws of a clamp-on ammeter must surround only one conductor. Including multiple conductors causes inaccurate readings because the magnetic fields of currents in opposite directions cancel. If the load to be tested is plugged into a receptacle, a line splitter is used. A *line splitter* is a plug-in device connected between a plug-in load and a receptacle that separates current-carrying conductors.

Thermal Inspections

Infrared thermometers, particularly thermal imagers, are extremely useful for a variety of electrical system inspections. Thermal inspections may not always reveal the cause of a problem, but they help identify issues that require further inspection with other instruments. Also, since conducting a thermal inspection does not require physical contact, it is a very safe way to assess a large amount of electrical equipment quickly.

For electrical systems, infrared thermometers are used to inspect fuses, circuit breakers, switchgear, transformers, motors, and many other types of electrical equipment, loads, and panels. **See Figure 5-12.** A concentrated area of higher than normal temperature indicates a problem. A certain amount of heat is normal, so for the most meaningful results, similar equipment under similar loads are compared. For example, when inspecting a three-phase fuse panel, the temperatures of components in each phase should be very similar. If component temperature on one phase is very different, then current on that phase should be measured.

Since most causes of motor failure are related to heat, a thermal inspection is a quick and effective way to evaluate motor health and efficiency. Thermal inspections may reveal both electrical problems, such as poor power quality, and mechanical problems, such as bad bearings.

Thermal Inspections

Fluke Corporation

Figure 5-12. Infrared thermometers, particularly thermal imagers, are often used to test electrical connections and motors for overheating.

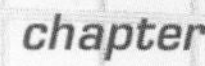

chapter 6

Auditing Compressed Air Systems

Electricity is used to operate the compressors that provide high-pressure air, which is then used as a power source or for other applications. For some industrial facilities, compressed air systems can use a significant amount of electricity. Auditing these systems primarily involves looking for air leaks.

COMPRESSED AIR SYSTEMS

A *compressed air system* is a network of piping, valves, and other fittings that distributes high-pressure air from a compressor to areas where the air will be used. **See Figure 6-1.** Compressed air systems are often used in industrial facilities as a power source or for applications such as cooling or cleaning. A compressor and its accessories must reliably supply clean, dry air for the system. Compressed air systems may account for approximately 10% of the electricity consumed in an industrial facility.

When used as a power source, compressed air systems are also known as pneumatic systems. Pneumatic tools use actuators to convert the flow of air into motion, either in a rotary or linear manner. This motion is then used to perform work, such as moving piston controls, powering hammer drills, or driving nails.

Compressors

An *air compressor* is a motor-driven device that intakes atmospheric air and forces it into a closed distribution system at a higher pressure. There are many different compressor types and sizes, which affect the operating efficiency and the quantity, pressure, and smoothness of the resulting airflow.

Air entering a compressor must be filtered and should be as cool as possible. Cool air is more easily compressed than warm air, so it improves compressor efficiency. Outlet air is always warmer than inlet air because compression raises the air temperature. The compressed air is then cooled, which condenses any water in the air so that it can be drained away.

The storage of compressed air allows the compressor to operate only intermittently while the air is used continuously. A *receiver* is an air tank that stores compressed air. Compressors and receivers must be sized to provide adequate pressure and to all the pneumatic loads while minimizing compressor energy use.

Fluke Corporation

Ultrasonic leak detectors are used to listen for the sounds of escaping air.

Compressed Air Systems

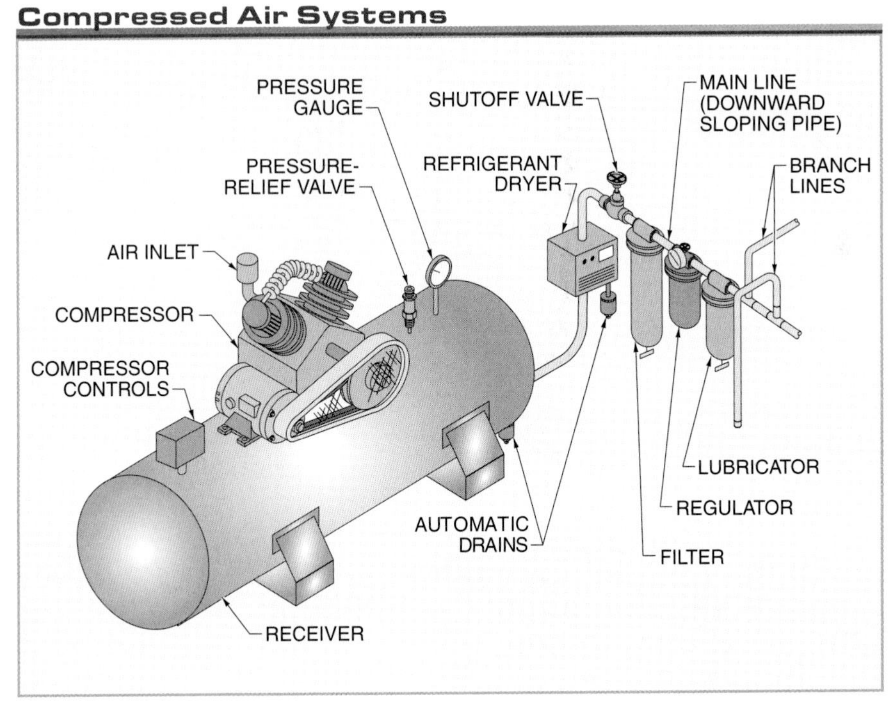

Figure 6-1. Many system components are required to produce, store, and distribute compressed air.

TECH-TIP

Compressed air as a power source has several advantages. The advantages include simplicity of design, low-maintenance tools, on-site energy storage, and increased safety in some circumstances, such as explosive environments.

Most compressors are driven by large electric motors. A simple motor control type is one that operates the compressor motor intermittently at full power in order to maintain the system air pressure within a range. **See Figure 6-2.** When the pressure drops to a low-pressure setpoint, the compressor starts up again to increase the pressure. When the high-pressure setpoint is reached, the compressor shuts down. Alternatively, compressor motors can be controlled with variable-frequency drives, which can adjust motor speed as needed to maintain a more constant air pressure.

Compressed Air Distribution Components

A compressed air system is composed of piping, fittings, valves, regulators, and other components that control, condition, and distribute compressed air to where it is needed. Due to the large number of connections and joints involved in assembling this type of system, it is the most likely source of leaks.

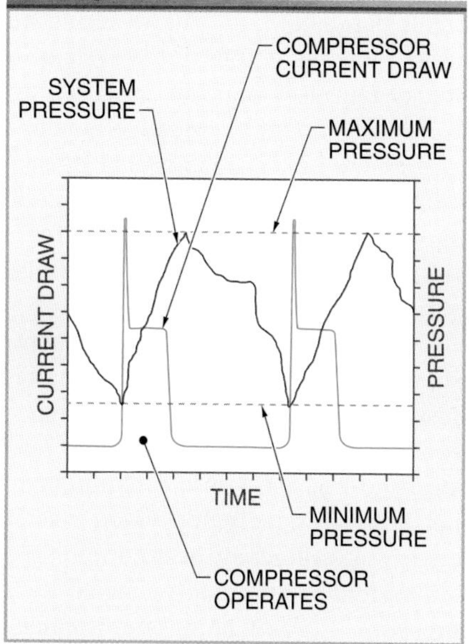

Figure 6-2. A compressor is operated periodically to maintain system pressure between two setpoints.

Leaks at connection points are often present at installation. Connections that are threaded may have leaks if not fully threaded or if the proper thread-sealing compound was not used. Connections that are soldered may have leaks if the solder does not completely fill the joint.

Rigid piping that is unsecured and subject to repeated motion may also develop leaks over time. If a location does not allow rigid piping to be secured against a permanent structure, flexible hose may be an appropriate alternative.

Valves and other components with moving internal parts may become damaged from corrosion due to moisture in the air or debris from unfiltered air. The damage may cause gaps between sealed surfaces, resulting in leaks.

AIR LEAKS

Air that leaks from a compressed air system escapes the system unused. Since this air does no useful work, the energy used to compress the air is wasted. Leaks also cause the system pressure to drop. This can result in the inefficient operation of pneumatic tools. The continuous discharge of air through leaks forces the compressor to operate more frequently, increasing maintenance requirements and shortening its effective service life.

Through all of these means, leaks can be significant sources of waste in industrial facilities. However, the most measurable of these is energy waste due to excessive compressor operation. Given the compressed air requirements of pneumatic tools, the required compressor operating conditions and its electrical power consumption can be determined. If the compressor uses more energy than expected, there may be leaks in the system.

Leaks can be located within sections of continuous piping when caused by damage or deterioration. For example, pipes can crack when crushed by moving equipment or develop holes when subjected to corrosive chemicals. However, most leaks occur at connections to fittings, such as couplings and valves. **See Figure 6-3.** Leaks often go unnoticed for long periods because air is invisible and piping may be hidden or inaccessible.

Improper installation or gradual deterioration can allow air to escape from between mated sections. In the case of quick-connect fittings, the design of the connection sacrifices some ability to hold pressure for the advantage of ease-of-use. Also, flexible hose material is easily damaged and can develop a gradual leak.

System Connections

Figure 6-3. The connections of compressed air system components, such as threaded joints, are likely sources of leaks.

COMPRESSED AIR SYSTEM MEASUREMENTS

The overall goal for auditing a compressed air system is to make the system meet the compressed air demand with less compressor operation. This is done by identifying air leaks and compressor performance issues. Baseline measurements are used to establish the current operating conditions and determine if there are significant leaks. Further tests are used to identify leak locations for repair.

Baseline Measurements

Baseline measurements establish how much energy is required to produce compressed air under normal operating conditions. The electrical energy consumption by the compressor is the input parameter, and the resulting air pressure is the output of the system. If the system uses multiple air compressors, each should be fitted with meters. A pressure meter is connected to each compressor output. These parameters should be logged continuously for a full day or longer if compressed air use varies daily.

The system pressure measurements should be compared with the requirements of the various loads. The compressor should meet the needs of the load with the highest pressure requirement but not excessively so, which wastes energy. Inefficiency in a compressed air system is commonly caused by a pressure setpoint being significantly higher than necessary, which requires the compressor to operate longer.

Leak-Down Test

A leak-down test measures the amount of time required for a fully pressurized system to completely depressurize due to leaks only. This test can only be conducted when no compressed air loads are operating. Therefore, it may need to be scheduled during off-hours.

A pressure meter is connected to the system. All valves within the system are opened, any connected pneumatic equipment is turned OFF, and any valves leading to unconnected lines are closed, which creates a single, pressurized volume. **See Figure 6-4.** The entire distribution system is pressurized to the normal operating pressure and then the compressor is shut off. Ideally, the system should hold its pressure for many hours or even days.

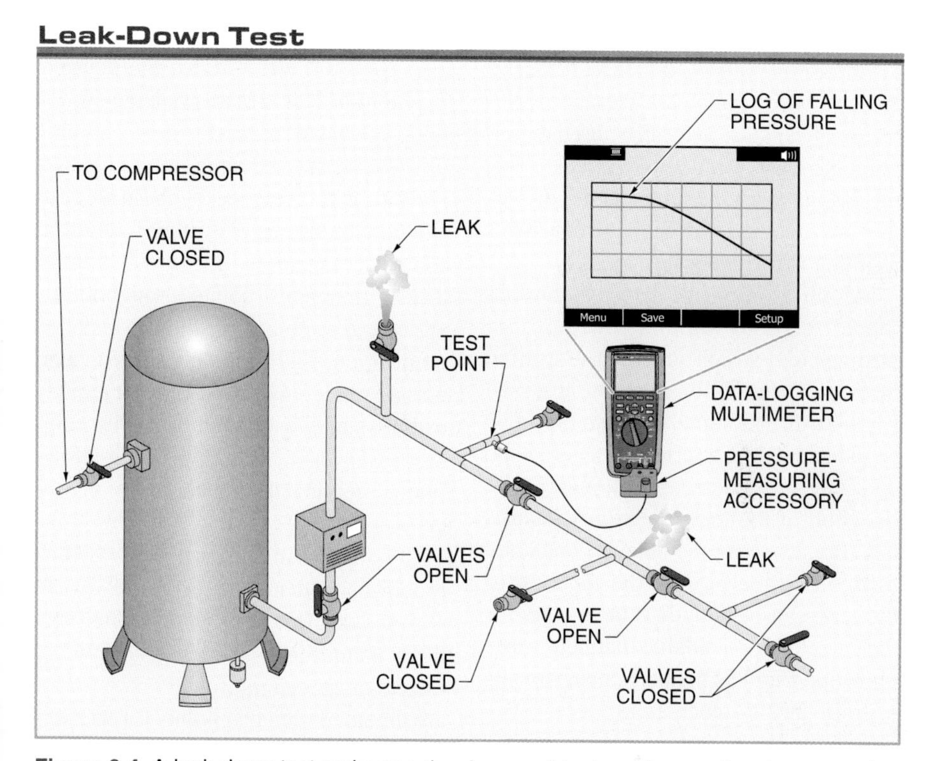

Figure 6-4. A leak-down test estimates the degree of leakage in a system by measuring the gradual loss of pressure.

The pressure meter is monitored as the pressure falls. If this occurs very slowly, a logging meter can be used to record the pressure over a long period. It should be documented how long it takes for the pressure to reach zero. The speed at which the system depressurizes indicates the degree of leakage.

Leak Location Detection

Once a leak problem is known, the location of the leak must be determined in order to remedy the problem. Auditors must also remember that the entire system should be checked, as there may be multiple leaks.

The simplest way to find leaks is through the application of soapy water to all piping and connections. The water is applied with a brush or sprayed on from a bottle. Escaping air will cause bubbles to form at the leak site. This method is generally effective but time consuming. Alternatively, test instruments can be used to quickly locate compressed air leaks.

Ultrasonic Testing. As air is forced through a small orifice, due to a break in the piping or loose fittings, the resulting turbulence generates pressure (sound) waves. Some leaks are audible within the range of human hearing and produce a hissing sound. These leaks are often found quickly by nearby operators.

However, many leaks produce sound in the ultrasonic range (20 kHz to 100 kHz) that cannot be heard. Ultrasound is very directional in nature and can be used to pinpoint the exact location of the leak. Also, the frequency of the ultrasound indicates the flow velocity, with higher frequencies resulting from higher flow rates.

An ultrasonic leak detector converts the ultrasound signals into audible frequencies, which are then amplified through speakers or headphones. **See Figure 6-5.** The signal strength may also be indicated with a series of LEDs on the detector. As the detector is pointed in the direction of the leak, the signal intensifies. By sweeping the detector back and forth along the piping, the location of the leak can be found. Attachments for the detector are available that help sense ultrasound signals in noisy environments or from hard-to-reach areas.

Ultrasonic Testing

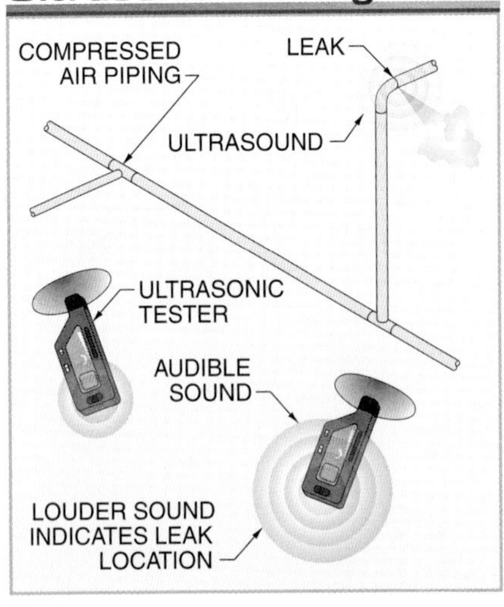

Figure 6-5. An ultrasonic tester is used to locate a leak, which produces a high-frequency ultrasound as air escapes the system.

Pressure Testing. Pressure testing is also used to narrow down the possible locations of leaks. In a very large system with extensive or hard-to-reach piping, this testing may be done first to identify sections with likely leaks. These sections can then be inspected further with ultrasonic leak detectors. Pressure testing for leaks requires test points throughout the system. Test points are fittings installed onto piping that include ports for temporarily attaching sensors. If not already installed, adding test points can be a significant investment in time, but it allows future pressure testing to be conducted very easily.

By measuring pressure at various locations and noting pressure drops, sections of piping can be identified that likely include leaks. Stand-alone pressure meters or special multimeter accessories are threaded onto test points for measuring pressure or vacuum.

Before taking a measurement, the meter needs to be zeroed. This establishes atmospheric pressure as the zero point on the measurement scale, which means that values will be in gauge pressure. Alternatively, absolute pressure measurements are relative to a perfect vacuum and have different numerical values for the same pressure. Auditors must understand the difference between these scales because a variety of pressure measurements are collected for an audit.

When the pressure transducer of the meter is connected to the test point in the system, the pressure reading is displayed. Multimeter accessories output a signal that is read by the multimeter as a small voltage with the same numerical value as the measured pressure.

chapter 7

Auditing Steam Systems

Steam systems are used to transport and distribute large amounts of heat energy, often for heating occupied spaces, but also for use in industrial processes. The large networks of pipes, valves, and other fittings are possible sources where some heat energy may be lost.

STEAM SYSTEMS

A *boiler* is a closed metal container in which water is heated to produce steam or heated water. Related equipment is used to safely distribute and control the steam or hot water for its intended application, and then return the hot water to the boiler to be heated again. **See Figure 7-1.** Steam is commonly used for industrial heating or process applications. For example, steam is used in food processing plants for cooking, pasteurization, and sterilization; in paper mills for pulp processing and drying operations; and in oil refineries for distillation of petroleum products. Hot water produced in a boiler is primarily used for building space heating applications.

Boilers

Boiler designs and sizes vary to accommodate a wide range of applications. Boiler designs are broadly classified as firetube and watertube boilers. A *firetube boiler* is a boiler in which hot gases of combustion pass through tubes surrounded by water. A *watertube boiler* is a boiler in which water passes through tubes surrounded by hot gases of combustion. Boilers are also classified as low-pressure or high-pressure types. Industrial process loads generally require the use of high-pressure boilers, which have a maximum allowable working pressure of 75 psi to 700 psi.

A boiler requires a feedwater system, fuel system, and draft system to operate properly. The feedwater system supplies water to the boiler at the proper temperature and pressure. Feedwater is treated and regulated automatically to meet the demand for steam. The fuel system supplies fuel in the proper amount to the furnace of the boiler where it is burned in order to change water to steam. Fuels commonly used are fossil fuels, such as natural gas, fuel oil, and coal. The draft system regulates the flow of air into and out of the boiler in order to maximize fuel combustion efficiency and discharge the gases of combustion out of the furnace.

Cleaver-Brooks

Boilers produce the steam that is distributed to heat exchangers throughout an industrial facility.

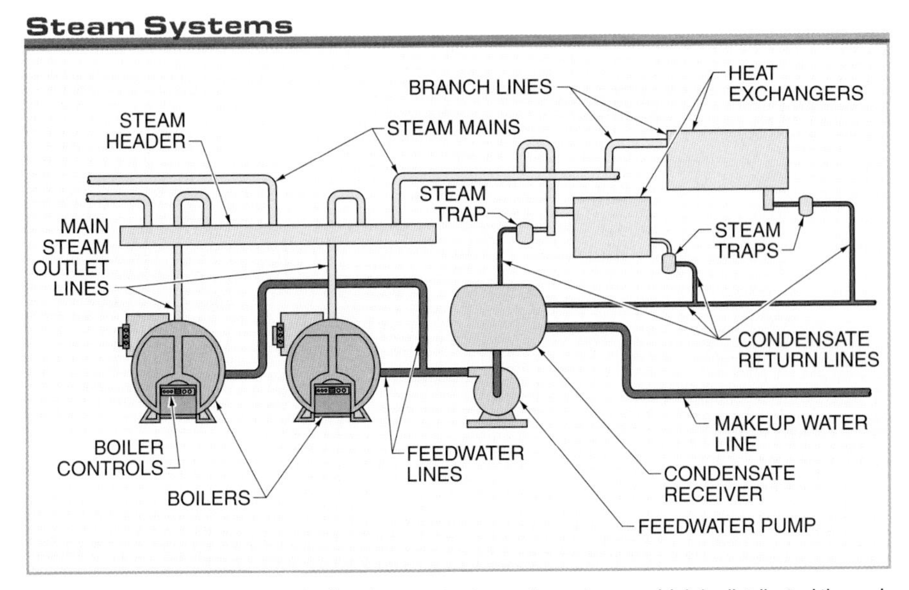

Figure 7-1. In steam systems, boilers heat water to produce steam, which is distributed through a network of steam lines and fittings to heat exchangers where the heat energy is used. The resulting lower temperature condensate is returned to boilers to be heated again.

Boiler operation is managed through electronic controls. A microcomputer runs a program that receives status information from various sensors and controls the fuel supply to the furnace and ignition.

Steam Distribution Components

A steam system collects, controls, and distributes the steam produced in a boiler. As steam heat is used, the steam condenses to water, which is recycled back to the boiler through the feedwater system. Steam is directed through piping to the point of use. A system of steam piping is made from steel and constructed of pipe and various fittings. There are multiple methods for joining pipe and fittings. **See Figure 7-2.**

Various accessories are needed throughout the steam system to control steam flow, maintain the required steam pressure, and remove water and air. Steam flow and pressure are regulated with valves and monitored with steam pressure gauges. Some valves are used to start or stop steam flow, and others are able to regulate the amount of steam flow.

A *steam trap* is a boiler accessory that removes air and condensate from steam lines without the loss of steam. **See Figure 7-3.** Steam traps are located downstream of wherever steam releases its heat and condenses, such as in a heat exchanger. The condensate collects in low parts of the piping, where the steam traps discharge the water to the condensate tank without releasing steam. Steam traps work automatically and increase boiler plant efficiency. They also prevent water hammer by expelling air and condensate from the steam lines.

Steam Pipes

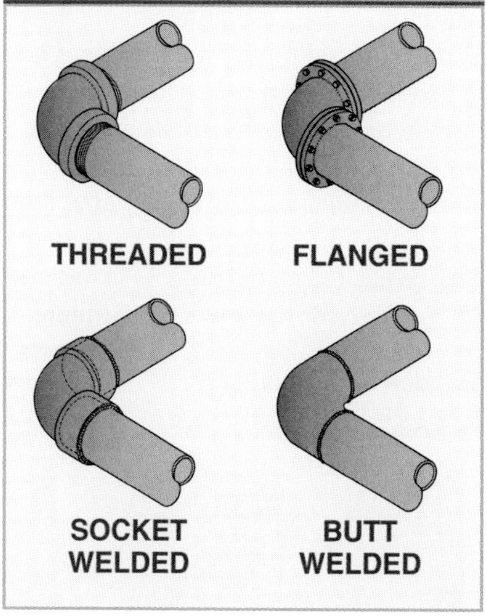

THREADED FLANGED

SOCKET WELDED BUTT WELDED

Figure 7-2. Steam pipes and fittings may be joined in a variety of ways. All connections are vulnerable to leaks that allow steam to escape unused.

A *steam strainer* is a boiler accessory that removes scale or dirt from steam. A steam strainer is located on the inlet side of each steam trap because scale or dirt can clog their discharge orifices. Steam strainers must be cleaned regularly.

STEAM ENERGY LOSSES

The efficiency of boiler operation and fuel consumption can sometimes be improved by inspecting boiler controls. This may require the services of an expert and licensed boiler technician. However, a more likely cause of steam system inefficiency is steam or heat losses in the steam distribution system. This portion of the steam system is easily inspected by energy auditors.

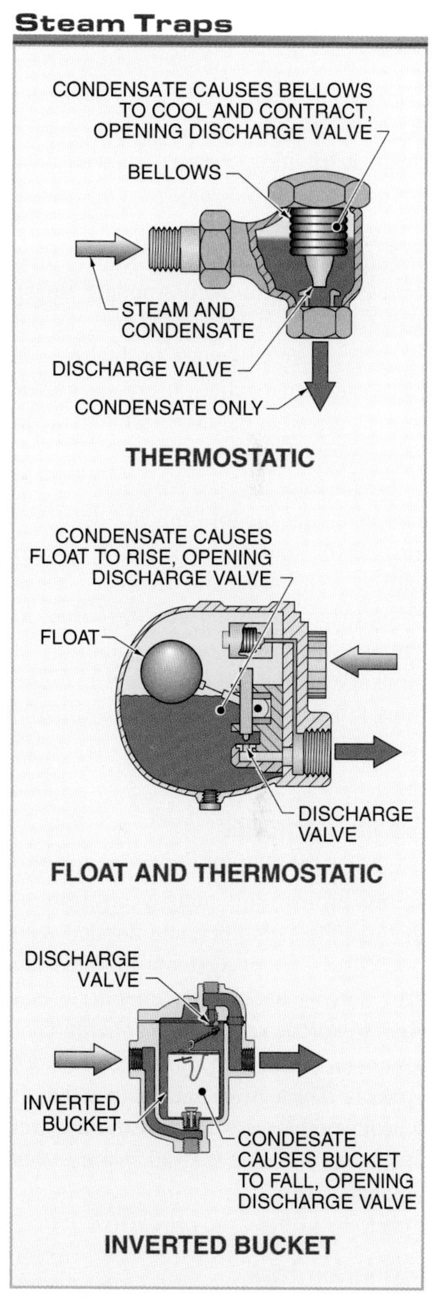

Figure 7-3. Steam traps separate liquid condensate from the steam system so that it can be returned to the boiler.

Steam Leaks

Any type of pipe connection is vulnerable to leaks, particularly when under the extreme temperatures and pressures of a steam distribution system. As a section of piping is heated by steam, the temperature change causes a significant expansion of the steel. Piping must be designed with measures to accommodate linear expansion of up to several inches, but the movement can still cause problems with connections. Installation errors, corrosion, or physical damage accelerates the occurrence of leaks.

Steam traps are another source of escaping steam. Steam traps are designed to remove condensate from the system without venting steam. However, these mechanical devices can fail due to damage, metal fatigue, or contaminant blockage of the small passages. If the traps fail in an open position, valuable steam is lost.

Steam Blockages

If a steam trap fails while closed, the steam piping and nearby equipment that use the steam become flooded with condensate. Clogged steam strainers can also cause a backup of condensate that similarly prevents the proper operation of steam distribution and condensate return systems. These blockages reduce the efficiency of the system or cause additional problems in the industrial process that uses the steam.

Insulation Loss

Steam distribution pipes are wrapped with insulation in order to minimize heat loss. **See Figure 7-4.** Some areas must remain uncovered, such as around certain fittings, but insulation should cover as much of the piping as possible. Lack of insulation in any area accelerates heat lost through the material of the steam pipe and fittings. Heat loss reduces steam temperature, which may adversely affect the process that uses the steam. More steam may be needed by the process in order to compensate for the lower temperature.

Steam System Insulation

Figure 7-4. Steam pipes and fittings are wrapped with thick layers of insulation wherever possible to minimize heat loss.

If excessive, the heat loss may cause the steam to condense back to water, which may obstruct steam flow. Also, the ambient air gains the heat lost by the steam. If the piping passes through conditioned spaces, it may add cooling loads to the HVAC system.

STEAM SYSTEM AUDITS

Steam system audits typically focus on finding any steam or heat losses and verifying the correct operation of steam traps and strainers. Steam losses are leaks where steam actually escapes the system. Heat losses are excessive reductions in steam temperature (before the steam reaches its point of use) due to inadequate pipe insulation.

Ultrasonic Leak Detection

Like compressed air leaks, steam leaks are detectable using an ultrasonic leak detector. Steam escaping through a small opening emits sound. Leaks that are audible within the range of human hearing produce a hissing sound. However, many leaks produce sound in the ultrasonic range (20 kHz to 100 kHz) that cannot be heard. An ultrasonic leak detector converts ultrasound signals into audible frequencies, which are then amplified through speakers or headphones. As the detector is pointed in the direction of the leak, the signal intensifies. By sweeping the detector back and forth along the piping, the location of the leak can be found.

Thermal Inspection

Due to the high temperature of steam and condensate, steam distribution systems are ideal subjects for inspection with infrared-based test instruments. Many areas can be easily checked with noncontact infrared thermometers, which provide individual temperature readouts of small areas. **See Figure 7-5.**

Thermal imagers provide temperature information across a wide field-of-view. Any steam loss or heat loss is shown as an abrupt increase in temperature. For example, escaping steam may be indicated by unexpected hot areas surrounding a steam pipe. All piping and fittings should be inspected to identify any steam leaks or blockages. Steam pipe insulation should be inspected for loose or missing sections. Loose or missing sections of piping insulation should appear on a thermal imager as warm areas. Thermal imagers can also be used to inspect insulation on the outer surface of boilers.

Measuring Steam Pipe Temperature

Fluke Corporation

Figure 7-5. Infrared thermometers are used to make spot temperature measurements of steam system components or piping.

TECH-TIP

A strainer is used upstream of a steam trap because the orifice through which the condensate flows inside most steam traps is quite small and easily clogged with debris. A strainer is a filter used to catch these impurities before they can plug the orifice but allows steam and condensate to pass.

Old insulation may contain asbestos products that require special handling procedures. Unless the auditor is certain that no asbestos or other harmful materials were used, the insulation should not be touched or disturbed in any way during the audit. Areas where steam leaks, steam obstructions, or insulation problems are indicated should be noted for appropriate inspection later while using the required PPE.

Steam traps that are operating normally should have a large temperature difference between their inlets and outlets. **See Figure 7-6.** A relatively small temperature difference indicates that the trap has failed in the open position, and steam has been allowed to pass through to the condensate return system. The temperatures of both sides of each valve can also be compared to confirm valve position.

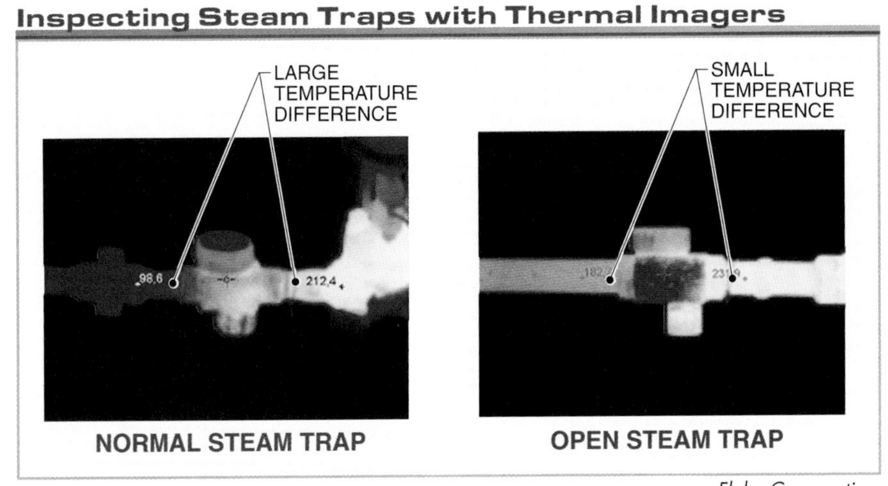

Figure 7-6. Thermal imagers display the temperature differences across a steam trap as color changes, which indicate whether the steam trap is operating normally.

chapter 8

Auditing Water Systems

Unlike most building systems, which rely primarily on electrical or fuel energy, water systems are usually not a significant consumer of energy. However, water is a resource that is supplied to a facility at a cost, meaning that identifying wastes and conserving the resource reduces costs. For this reason, water systems are sometimes included in energy audits.

WATER SYSTEMS

Nearly every building includes a water system for plumbing and basic cleaning needs. The costs associated with this type of water use are typically small, which means that these systems may not be included in an audit plan. However, many commercial buildings and industrial facilities use large amounts of water for special purposes, such as in manufacturing processes, large-scale equipment cleaning, or landscape irrigation. In these circumstances, it may be cost-effective to audit a water system for savings opportunities. If the water use of a facility is large, then a reduction in water use may provide significant cost savings.

Water is itself a utility resource with monetary costs. A water system may or may not also involve appreciable amounts of energy consumption. Some water systems include electrically operated pumping, filtering, and treatment equipment, particularly if the water source is on-site, such as a pond or well. Water supplied from a municipal supply may still require pumping if the pressure must be increased. In these cases, the electrical energy use associated with pumping or treating the water may deserve an audit inspection.

Water system design and components may vary greatly between facilities, depending on the end uses. Like compressed air and steam systems, water systems generally consist of a network of piping that includes valves, gauges, connectors, and other fittings for controlling the water pressure and flow. **See Figure 8-1.** The pipes may run through walls and floors, behind equipment, overhead, or outdoors underground. Pumps and related equipment are typically located in basement or utility areas so that they are reasonably accessible.

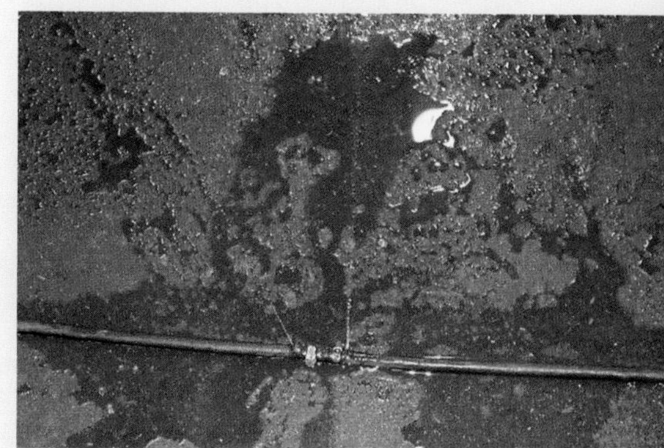

Fluke Corporation

Water leaks are not only wasteful but also may damage equipment or create safety hazards.

Water Systems

Figure 8-1. Water systems consist of piping, fittings, valves, pumps, and other equipment for distributing large amounts of water in industrial facilities.

Another source of water-related utility costs is wastewater. Municipal wastewater disposal, through the sewer system, is a billed utility service, typically based on water usage. On-site disposal often involves treatment and/or monitoring to ensure no adverse environmental effects. Reducing water use also reduces the amount of wastewater disposal, so water system audit cost savings are increased further.

WATER LEAKS

Leaks allow a resource to escape a system unused, requiring more of the resource to be delivered so that the end use has an adequate supply. Like compressed air and steam distribution systems, water leaks waste a resource that has measurable financial value to the operation of the facility.

Also, water leaks can cause other problems more costly or harmful than the cost of the water itself. Water infiltration can permanently damage equipment through corrosion, electrical shorts, and interference with the lubrication of moving parts. Wet floors cause slip and fall hazards. Pooled water near electrical equipment creates a shock hazard. Stagnant moisture encourages mold growth and degrades porous materials.

Water leaks are commonly found at junctions between pipes and fittings. The leak is often caused by improper installation, corrosion, vibration, or physical damage. Another source of leaks is the use of flexible hoses. These hoses and their connectors are easily cut or deformed unintentionally, especially in an industrial setting.

TECH-TIP

If water is heated on a large scale for industrial use, the equipment used to raise the temperature is another potential source of energy waste. At this scale, a hot water boiler is likely used, and the hot water distribution system should be audited similarly to a steam distribution system.

WATER SYSTEM AUDITS

Audits of water systems should trace all industrial water piping from the source to the point of use when inspecting for leaks. If there were a known discrepancy between the water supplied at the source and the water used at the end points, significant water leaks would already be suspected.

Auditors should also discuss the audit with the personnel working in the audit areas. Personnel may already be aware of some leaks, but if the leaks are not considered significant or hazardous, they may have been ignored.

Baseline Water Use

Part of an audit preplanning process includes collecting all utility bills from the previous year or more. Utility bills provide energy or resource use on a monthly basis. This information is used to establish baseline consumption trends, including average use and any seasonal variations. Estimates of water savings, and later actual water savings, are compared to the baseline information to evaluate the priority and effectiveness of conservation measures. **See Figure 8-2.**

Baseline Water Use

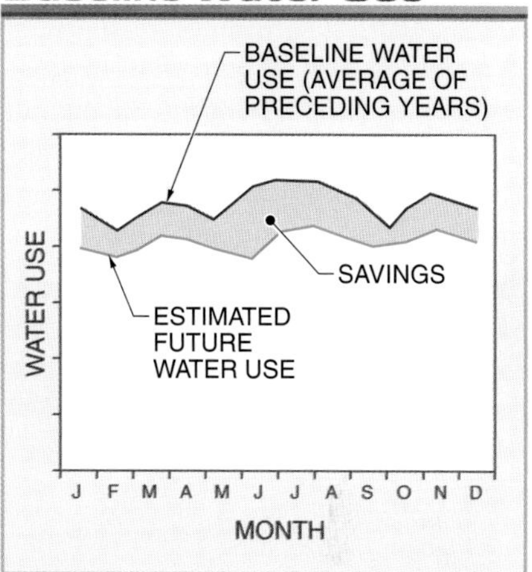

Figure 8-2. Auditing a water system begins with establishing baseline water use based on the utility bills of preceding years.

Visual and Auditory Inspections

Water leaks are often relatively simple to identify visually. Most leaks cause obvious puddles on floors or equipment. Puddles form on the lowest surfaces, however, and may not be directly under the leak. Trails of water can be traced back to the source to find the leak. Flashlights or portable lighting can be used to identify wet spots on some surfaces, which shine in the light. Reducing the normal area lighting to low levels may help make the shine more visible.

The most difficult aspect of auditing water systems is inspecting piping that runs inside of walls, behind equipment, or in otherwise poorly accessible spaces. In these cases, other signs of water problems may be helpful in locating a leak. For example, high-pressure leaks may produce audible hissing sounds, while low-pressure leaks may produce dripping sounds.

Thermal Inspections

Thermal imagers may identify water or moisture problems in some situations. **See Figure 8-3.** If the water is hotter or colder than the surroundings, the areas around water leaks should appear noticeably different in the infrared image. If the water is not heated or chilled, it may still appear slightly cooler than the surroundings in the image. Unless the ambient humidity is very high, the evaporation of the water makes it appear cooler.

However, conditions may not be conducive to identifying moisture with a thermal imager. A lack of moisture signatures from a thermal inspection does not guarantee the absence of leaks. Although, if signs of moisture problems are present, thermal imaging provides an easy method for quickly screening large areas so that suspected problems can then be inspected more carefully.

TECH-TIP

Thermal imagers are particularly useful for conducting energy audits because they can detect several different types of energy waste, are easy to use, and can scan an area quickly.

Thermal Inspection of Water Leaks

Fluke Corporation

Figure 8-3. Thermal imagers may detect water or moisture and display them as areas of a different color.

chapter

Auditing Waste and Recycling Programs

Auditing the waste and recycling programs at a facility primarily involves reviewing documentation and planning improvements for waste management procedures. Cost reductions are achieved by reducing the production of waste within the facility and maximizing the diversion of recyclable materials into income-producing programs.

WASTE AND RECYCLING PROGRAMS

Waste services and recycling programs are not energy-related systems, but they incur costs associated with facility operations that can potentially be reduced through an auditing process. Reducing the amount of waste requiring disposal also provides environmental benefits.

Waste Disposal

Most industrial waste is composed of ordinary, nonhazardous materials that are eventually disposed of in a landfill. The most common types usually come from leftover raw materials, defective products, and discarded packaging. Therefore, the actual materials involved depend on the type of facility and the processes used. For example, a facility that manufactures household appliances is likely to discard materials such as sheet metal, plastics, wiring, and cardboard.

Many industrial facilities also produce waste that is considered hazardous because of its toxicity, flammability, corrosivity, or reactivity. Examples include chemicals, used oil, batteries, and fluorescent lamps. This waste must be handled with special equipment and by trained personnel. **See Figure 9-1.**

Hazardous Waste Handling

Mine Safety Appliances Co.

Figure 9-1. Hazardous waste must be handled, stored, and transported under special procedures prior to disposal.

Large commercial and industrial facilities have their own waste management procedures for promptly collecting, transporting, and sorting waste from throughout the facility and storing it temporarily. The waste is then periodically removed by waste disposal contractors. Waste contractors either empty the waste containers into their trucks or replace full containers with new, empty ones. These waste removal services are typically billed according to the size of the container and the frequency of removal.

Recycling Services

Recycling is the processing of used materials into a form that can be remanufactured into new products. The role of a consumer, including industrial facilities, in recycling involves the separation and sorting of recyclable materials, which are then collected periodically in a way similar to waste removal. Only certain materials can be recycled efficiently, including glass, metals, plastics, paper, cardboard, and textiles.

Waste recycling provides multiple benefits. First, it diverts waste that would otherwise be disposed of in a landfill or incinerated at a waste facility. This both reduces waste removal costs and is an environmental benefit. Also, since the costs of reprocessing some recycled materials are less than the costs of producing more of the material from its raw constituents, recycling is a cost-effective manufacturing process. **See Figure 9-2.** Therefore, recycled materials have an intrinsic value that makes them an asset to the facility.

Recycling services often are offered by waste contractors, but they may also be provided by specialty recyclers. The materials are brought to sorting centers where materials are cleaned and separated by alloy (for metals), polymer type (for plastics), color (for glass and plastics), paper type (for papers and cardboard), and other criteria. Each type of material is then sold to the appropriate reprocessing plant for conversion into manufacture-ready materials.

Any nonrecyclable materials are separated and disposed of as ordinary waste. If the mixed recyclables include an unacceptable amount of nonrecyclable materials, the recycler will likely charge fees to the origin facility, which may offset the income received.

WASTE COSTS

Waste is a cost because there are expenses involved in removing and disposing of these materials. Excessive waste is also a symptom of process inefficiencies.

Recycling

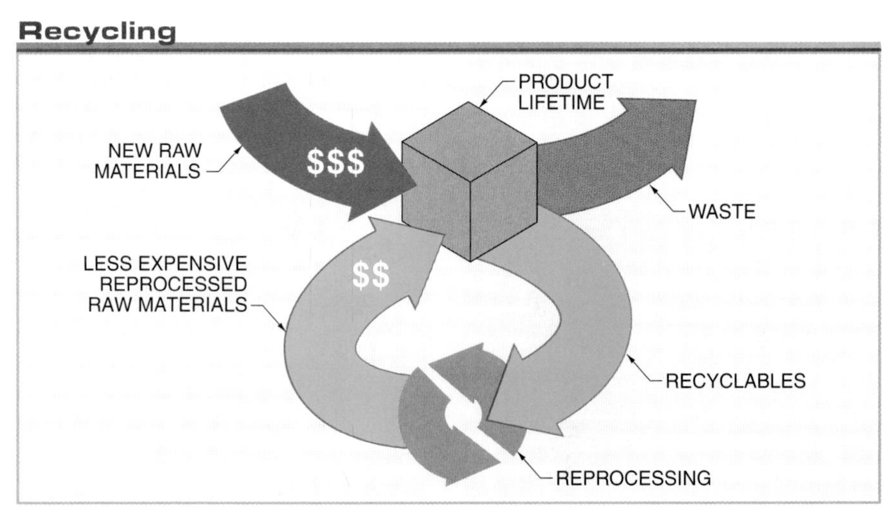

Figure 9-2. It often costs less to reprocess recyclable materials than to make original raw materials manufacturing-ready. Therefore, recyclables are assets with income potential.

Excessive Waste

Nearly every facility produces waste that must be disposed of, but the amount of waste may be more than necessary. Since waste disposal incurs costs, failure to minimize the disposal needs can be considered an unnecessary expense that could be reduced through auditing.

An excessive accumulation of waste is likely due to inefficient processes and/or underutilized recycling opportunities. Process-related waste streams include leftover raw materials, product packaging, old equipment, used maintenance supplies, and office waste. Inefficiencies in nearly all aspects of industrial processes may contribute to material waste.

Recycling programs not only divert a portion of the waste stream into less expensive or cost-free disposal, but they may even provide an income. **See Figure 9-3.** Therefore, neglecting to maximize recycling participation incurs a cost penalty.

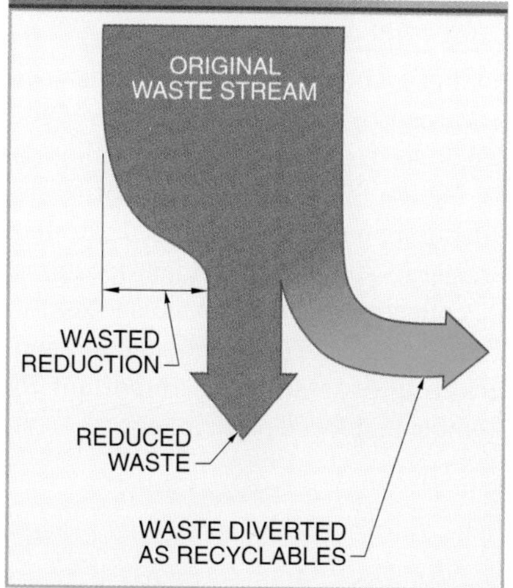

Figure 9-3. Excessive waste can be reduced by implementing process-related changes and/or diverting recyclable materials.

Surplus Waste Capacity

Waste disposal services are unique in that the costs are not always tied to exactly how much is removed. Contracts may specify the removal of waste up to a certain amount (such as the volume of an on-site waste container) at regular intervals. The cost per interval usually remains the same, regardless of whether the maximum amount was removed. Therefore, if waste can be reduced to a level that is consistently below the current capacity, there is a cost saving opportunity in renegotiating for a smaller container size.

WASTE AND RECYCLING AUDITS

Waste and recycling audits are relatively simple to perform, so it is recommended that they be conducted annually regardless of when other energy audits are repeated. Audits should consider all waste streams, including ordinary waste, recyclables, and special or hazardous waste.

Waste Production

Auditors should evaluate the sources of significant waste streams and investigate whether changes to purchasing, handling, processes, or procedures can reduce the production of waste in the facility. For example, if discarded packaging materials are a significant source of waste, a review of alternative suppliers may yield a comparable product that uses less packaging. Also, ways to reuse potential waste should be sought to provide an added benefit or displace another cost. For example, shredded office paper can be used as packing material.

Handling and Sorting Procedures

If a recycler places a higher value on presorted recyclables, then it may be cost effective to incorporate sorting into the in-house waste collection. Auditors should investigate the equipment, procedures, signage, and training involved to implement a comprehensive recycling program. Recyclers may provide guidelines on setting up a recycling program for maximizing payback.

It is very important to keep material separate throughout the sorting and collection process. Even a small amount of contamination or mixing of recyclables can turn these income generators into expenses. This is because the recyclable material will either require extra sorting or be treated as ordinary waste and be disposed of through the normal waste channels.

Invoice and Contract Reviews

Auditing waste and recycling programs and procedures begins with reviewing the current service contracts and past billing. The services, disposal minimums, surcharges, and fees should be noted and compared to the terms offered by competing contractors. Breakdowns of charges should be requested so that services can be compared appropriately.

Multiyear service contracts may have low initial fees but lock the facility into a service that does not allow renegotiation. As the facility undergoes a comprehensive waste reduction campaign, waste and recycling contracts should be kept to one-year intervals so that the services and terms can be renegotiated annually for reduced disposal requirements.

Equipment Inspections

A facility with a large amount of one type of waste may utilize equipment to compact and store nonperishable materials. Compactors are particularly common for flattening and bundling cardboard and textile wastes. These machines are typically hydraulically powered to compress the material, which maximizes the use of the container capacity and reduces the frequency of removal. Like any mechanical equipment, compactors should be regularly inspected for proper operation, lubrication, hydraulic pressure, and structural integrity.

Personnel Survey

Auditors should interview key personnel in the waste management and production areas of a facility. These employees are likely to have suggestions for waste reduction and recycling procedures and be able to identify convenient locations for collection points. For the success of a new or expanded waste reduction and recycling program, it is important to promote employee participation.

Auditing Plug-In Loads

Plug-in loads are not associated with specific building systems because they may include such a wide range of devices. Any plug-in load can potentially waste energy. However, the plug-in loads that are most likely to waste energy include computers, office equipment, and other electronic devices.

PLUG-IN LOADS

In an industrial facility, the most significant electrical loads are likely to be hardwired into dedicated panels. A *hardwired load* is an electrical load that cannot be disconnected from the electrical system without de-energizing a portion of the system. However, there are still receptacle branch circuits wired throughout the building to accommodate small, portable, temporary, or unanticipated loads that are plugged-in as needed.

A *plug-in load* is an electrical load that is connected to the electrical system through a receptacle. Receptacles and the plugs of plug-in loads are easily connected and disconnected with minimal exposure of live electrical parts. They also do not require any part of the electrical distribution system to be de-energized. **See Figure 10-1.** A number of receptacles can be wired together in parallel as a single branch circuit.

Plug-in loads are typically portable devices that may be moved between locations in a building. Examples of plug-in loads include portable lights, fans, computers, printers, copiers, televisions, clocks, and some appliances.

Industrial facilities may include a significant number of plug-in loads. For example, any area dedicated for office functions is likely to include many plug-in loads, such as computers and copiers. Commercial buildings are likely to include an even larger number of plug-in loads due to the types of work performed in the buildings. The portion of total energy consumption from plug-in loads varies considerably, but these loads should always be investigated during an energy audit.

Plug-In Loads

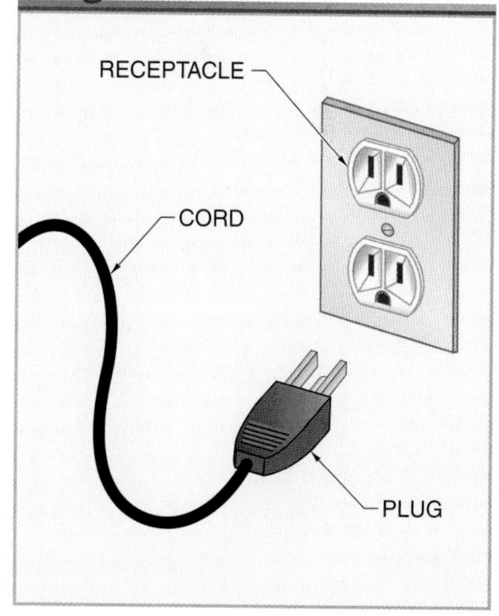

Figure 10-1. A plug-in load is any type of load that receives electrical power through a plug and receptacle arrangement. A flexible cord delivers power from the plug to the load.

UNNECESSARY POWER DRAW

In the course of an audit investigation, it may be discovered that plug-in loads are damaged, misused, or otherwise operating in a way that wastes energy. For example, a portable heater being used near an open doorway allows the heat to escape outside. Since the range of possible plug-in load types and circumstances are so broad, the scope of possible issues cannot be easily listed. An auditor's experience with other building systems and their common energy issues should be adequate to identify similar issues with plug-in loads.

However, unnecessary and avoidable power draw is a common cause of wasted electricity among electronic plug-in loads and should be investigated specifically. The plug-in loads are either operating when they are not needed, or they are somehow wasting energy even when in an inactive state.

User Operation

Energy for plug-in loads is mostly wasted when the loads are left ON when not needed, such as during nonworking hours. The primary cause of this is user habit, since plug-in loads are generally under manual control. It is more convenient for users to leave loads ON all the time rather than diligently switch them OFF when they are no longer needed.

TECH-TIP

DC power supplies use transformers and rectifiers to convert alternating current to direct current, typically at a far lower voltage. These components all create waste heat in the conversion process.

It is also believed that the additional energy consumption of idle equipment is less costly than the maintenance and replacement, due to wear-and-tear, of equipment that is shut down and restarted daily. However, there is little evidence to support this assumption. It is commonly asserted that for modern electronic equipment, the wear of daily restarting is negligible and less costly than the additional energy consumption.

Standby Power Draw

Standby power is the electrical power drawn by a load that is in a reduced-power, inactive state. When switched OFF, many electronic devices do not completely shut down but instead revert to a standby state. This allows the device to retain the current time or settings information, await "Power On" inputs from remote controls, or return to an active state very quickly.

Power supplies for DC loads may draw power in a way similar to standby power when the loads are not operating. Many plug-in loads, particularly electronics, operate internally from DC power. Since most facilities do not provide DC power distribution, a DC load must include a power supply component that converts the commonly available AC power to usable DC power. Some DC power supplies are located inside the load, like those for desktop computers. Many are separate units, such as the power supplies for small electronic devices. **See Figure 10-2.** The transformers inside DC power supplies usually draw power, although reduced, even when the load is turned OFF. Stand-alone power supplies may continue to draw power even if the load is disconnected.

External DC Power Supplies

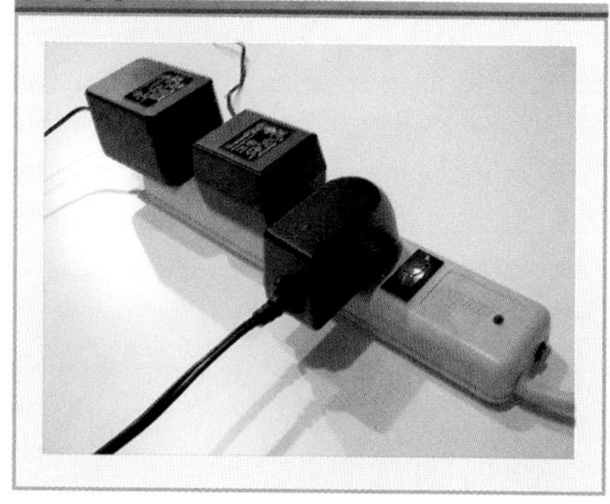

Figure 10-2. Electronic devices usually require a DC power supply, which may be an external component in the form of a very large plug.

While these types of power draw are relatively low, the energy consumed is significant with a large number of devices and usage over a long period of time. A common and simple method for addressing these issues is to plug the power supplies into switchable power strips. The power draw to all connected devices is completely stopped by switching the power strip OFF.

Heat Gain

DC power supplies are also sources of heat. For example, an external power supply is usually quite warm to the touch after hours of use. Discrete electronic components also generate heat. When concentrated inside of devices, the heat is great enough to require small internal fans to dissipate it before the components are damaged. Rooms with many electronic loads are often noticeably warm to occupants and require additional cooling.

The additional heat may be helpful during the winter months, but additional cooling is required the rest of the year. Better management of the operation of plug-in loads may be able to reduce the energy consumption of both the electronic loads and the HVAC system.

PLUG-IN LOAD AUDITS

Auditing plug-in loads involves carefully recording information and taking electrical and thermal measurements. These tasks identify plug-in loads that are operating unnecessarily, drawing power when inactive or OFF, or generating excessive amounts of waste heat.

Plug-In Load Inventory

A walk-through of the entire facility should be conducted, noting any significant plug-in loads. An inventory of these loads should include type, quantity, location, usage characteristics, and any other relevant information. Particular attention should be paid to electronic equipment, such as computer and office equipment. If any of these devices have an energy-saving mode, it should be noted.

Energy Consumption

For each plug-in load, the energy consumption should be determined. Many plug-in loads have their power draw listed on a label provided by the manufacturer. Daily energy consumption can be estimated by multiplying the power draw by the total time the load operates each day. This is convenient for active operation, but labels may not reveal standby power draw.

If power draw information is not available, or it is suspected that a load draws power while inactive, the current draw can be directly measured with a clamp-on ammeter. Line splitters are also commonly used to measure current draw. **See Figure 10-3.** The current draw multiplied by the circuit voltage equals the power draw. The current draw of an entire branch circuit, such as one feeding an office area, can be taken directly at an electrical panel. Measurements taken with all loads OFF or in an inactive state determines whether there is significant standby power draw.

Measuring Energy Consumption

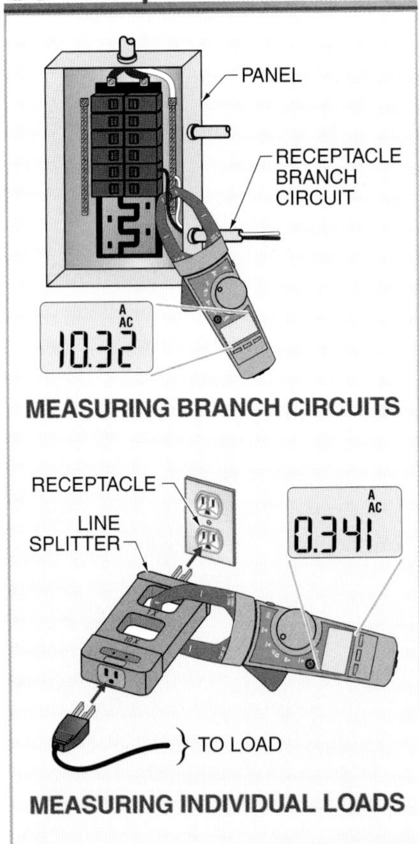

Figure 10-3. The energy consumption of plug-in loads can be determined by directly measuring the current draw.

Depending on the results of panel measurements, further measurements at power strips or individual loads may be necessary to identify particularly large draws of power. Dedicated energy consumption meters that plug in between the load and the receptacle can also be used.

Heat Detection

Hidden and undesirable sources of heat from plug-in loads are easily detected using noncontact infrared thermometers. **See Figure 10-4.** Thermal imagers are particularly useful because they allow a large area to be scanned quickly for hot spots. The images can then be stored as documentation. Besides plug-in loads themselves, auditors should also inspect receptacles, external power supplies, and electrical cords.

Measuring Waste Heat

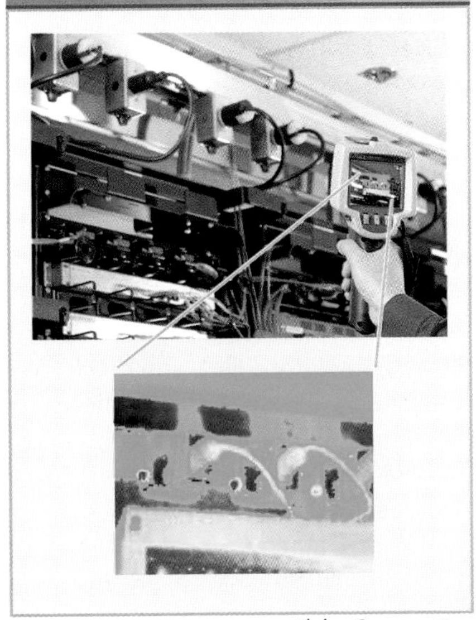

Fluke Corporation

Figure 10-4. Waste heat from plug-in loads is detected with infrared thermometers or thermal imagers.

chapter 11 Improving Energy Efficiency

The energy audit itself does nothing to reduce energy use or lower utility costs. The information gathered during the audit is used to formulate recommendations to change facility equipment, operations, or procedures in a manner that will reduce energy or resource use. The energy audit process is then used again to verify and sustain the efficiency measures. Ongoing monitoring and repeat auditing ensures continuing performance over time and identifies new opportunities for energy conservation.

DATA ANALYSIS

The data gathered from energy audit inspections, measurements, and data logging is analyzed to identify and quantify savings opportunities. Accurate data and realistic estimates are critical for maximizing the return on investment (ROI). The analysis of audit data is used to compile an audit report, which is then used to choose which projects will be implemented and plan their execution.

Software Tools

Software tools are useful for nearly all aspects of an energy audit. After performing inspections and system measurements, the data can be entered into software as an organized database. If consumption information over time is available, trends can be plotted and analyzed as graphs. Software is also useful for performing calculations quickly and accurately. Some software, when enough information is entered, can be used to automatically generate audit reports.

Specialized software is available that is designed particularly for energy audits. This type of software is especially helpful for first-time auditors because it likely includes hints on the necessary measurements to be taken. Much of this type of software is available for free online. Sources include the Department of Energy programs, state agencies, and some manufacturers. **See Figure 11-1.** Related online resources are also helpful for comparing the energy performance of the facility against similar facilities.

Software Tools

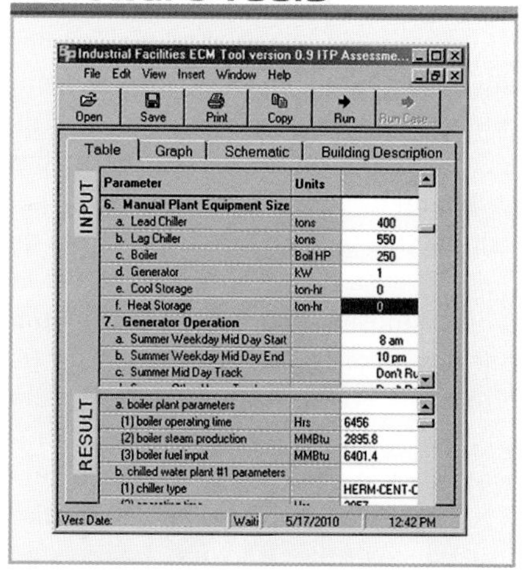

Figure 11-1. Various software tools are available that can be used to store, organize, analyze, and calculate energy audit data.

Alternatively, general-purpose word processing, database, and spreadsheet software can be used to organize audit data. This may work equally as well but requires someone experienced in energy audits to set up the files. An advantage of this method is that the data can be organized and presented in exactly the way desired.

Potential Savings

System efficiency may have decreased over time due to equipment age, wear and tear, poor maintenance, power quality problems, and physical damage. If past performance data is available, the equipment efficiency can be evaluated for long-term trends. Eliminating the source of energy waste or inefficiency represents a savings in the form of reduced costs.

Equipment specifications and measurements of energy waste are used to calculate potential savings. The exact method of calculation may vary, depending on the system and energy source. However, the basic method for calculating potential savings involves proposing a change to the system, estimating the energy use of the modified system, calculating the cost of that energy use, and subtracting the new, lower cost from the present cost. **See Figure 11-2.**

The most difficult part is estimating the new energy use information for the modified system. For system repairs, such as fixing leaks, the new energy use is based on the original (when new) specifications, with some accounting for age or other factors. For system upgrades or equipment replacements, the new energy use is based on the specifications for the new equipment. For some systems, such as HVAC systems, it may be necessary to consult with outside experts to quantify energy use estimates.

Estimated Potential Savings

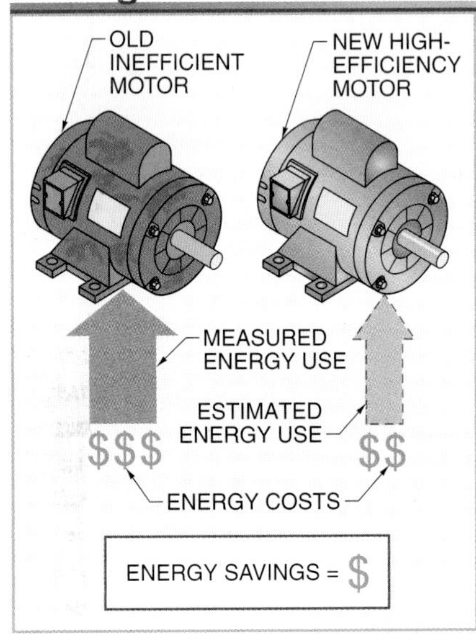

Figure 11-2. The potential savings from an energy efficiency project are estimated from information on the energy use of new or modified loads.

Incentives

Government entities, utilities, and nonprofit organizations sponsor incentives to make energy-efficiency and resource-conservation projects more affordable. An *incentive* is a monetary inducement to invest in a certain type of capital improvement that provides some societal or environmental benefit. The benefits to be received from incentives should be analyzed when calculating the ROI.

Energy-efficiency incentives encourage energy consumers to implement changes to systems or equipment that will reduce energy use. Incentive programs are available for offsetting the costs of energy audits, high-efficiency equipment, improved system controls, motor drives, better insulation and weather sealing, utilizing waste heat, and installing renewable energy systems. In the context of energy cost savings, financial incentives can usually be treated as an extra savings component.

Incentives may take many forms. One-time reimbursement payments include rebates and grants, which offset a portion of the initial investment. Loan programs provide low-interest or no-interest loans for the purchase of equipment, lowering the cost of borrowing money. Tax incentives lower owner tax liability, reducing future costs.

The availability of incentives varies widely by project type, state, utility, sector (residential, commercial, educational, etc.), and implementation. Many incentives are also time-limited, so information must be checked frequently for new programs and to ensure that an existing program has not expired.

Implementation Costs

Almost all energy-conservation measures will incur financial costs. Some may be very small, such as the implementation of new procedures, while others will be significant investments, such as replacing large equipment. In order to determine which of the possible projects are the most cost effective, the costs of implementing the programs must be determined.

The easiest types of projects to put a price on are those that require outside contractors. The contractors can be requested to prepare bids for the work. An open bidding process with several competing contractors is recommended in order to get the best value. The facility owners or managers should prepare a bid specification that lists exactly the work to be accomplished. If necessary, a list of possible options or alternatives is given. It is important that the bids be for the same tasks so that they can be compared appropriately.

Using outside contractors simplifies the implementation process but does not guarantee results. Unless the outside contractors were involved in the energy audit measurements, they are not likely to guarantee a certain energy savings resulting from the project. Therefore, the facility should consult closely with the contractors to clarify the expectations and obligations. The contractors may even suggest changes to the project plan that will increase the chances of maximizing savings.

For projects that are completed by in-house personnel, the cost estimation process is more involved. It is relatively easy to get quotes for equipment and materials, but it can be difficult to determine the necessary labor for each task. The labor costs should be included, even though the personnel are already employees.

A thorough knowledge of the capabilities of the maintenance department is critical for estimating labor. Analyzing the reports of any past projects with similar tasks can be extremely helpful. These reports should include information of any problems encountered and the total time and number of technicians required to complete the project.

Also, the recommendations must consider any ongoing costs that are caused by the project. For example, an equipment replacement project has significant upfront costs but may also require additional maintenance and employee training costs that recur each year.

Audit Recommendations

With the potential savings and implementation costs calculated, the rest of the audit analysis is relatively simple. Cost effectiveness is based on the ratio of savings to implementation cost. If the savings are greater than the costs to implement the project, then it makes sense to recommend the project. **See Figure 11-3.** If the savings are small compared to the implementation costs, then it may not be cost effective to do the project.

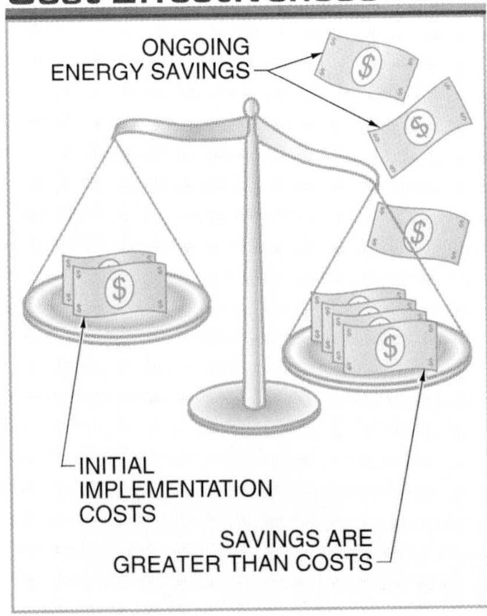

Figure 11-3. Cost effectiveness is determined by comparing the cost of a project to the savings resulting from the project.

Savings, however, are typically a periodic amount, such as $10,000 per year. Therefore, cost effectiveness is usually quantified as the length of time until the savings matches the cost. This is also known as the payback period. Further periodic savings after that point provide a net cumulative gain.

If the project is to be financed, the cost of borrowing money must be included. Also, if the ROI is expected to be more than about one or two years, it may be necessary to account for the changing value of money over time. These calculations are not difficult but require information on interest, inflation, and investment rates of return that must be researched carefully. A financial expert, such as an audit team member representing the accounting department, can be consulted to address this subject.

While financial considerations are the primary concern for choosing which projects to implement, other factors may also influence the decisions. These may include project requirements for safety, maintenance, employee training, physical space, and environmental measures. Any of these anticipated concerns should be included with the recommendations.

ENERGY AUDIT REPORTS

The energy audit report is the complete documentation of the energy audit and the results of the data analysis. The report is necessary for presenting the findings to those who must decide which projects to pursue and when. The goal of the report is to summarize all of the energy savings opportunities and produce a prioritized action plan and implementation schedule for addressing those opportunities.

The format of the report may vary, however, it must be easy to read and understand by the target audience. Visual components, such as photos, charts, and graphs, are especially helpful at conveying meaning. The critical summary information is often presented first, followed by detailed reference information.

Based on this information, items on the list should be prioritized. The easy-to-implement and high-ROI items are listed first. The savings available from implementing these projects first helps fund some of the others. This also builds momentum for the energy-efficiency program.

Project Summary

The most important part of an energy audit report is the summary of all the savings opportunities identified and their associated analyses. This information is typically presented in an abbreviated line-item format. **See Figure 11-4.** Each item represents a clearly defined source of energy waste or inefficiency. The important aspects to summarize are the recommended remedy, estimated savings, estimated implementation costs, realistic project duration, and ROI information.

Project Details

The bulk of the report is an expanded version of the summary. This section includes detailed explanations of the savings opportunities, supporting data, rationale for estimation methods, calculations, project tasks and schedule, a breakdown of implementation costs, and any other factors that must be considered. **See Figure 11-5.** It is also beneficial to designate a person responsible for each project, along with any other personnel that will be involved.

Audit Report Summaries

Priority	Project	Monthly Savings	Project Costs	Project Duration	Payback Period (Months)	Two-Year ROI
1	Add VFDs to air handler fans, recalibrate controls for variable volume	$4300	$10,000	2 months	2.5	$93,200
1	Raise summer temperature setpoint to 75°F	$2500	$2000	1 week	1	$58,000
2	Retrofit fixtures for fluorescent lamps	$2350	$27,000	2 months	11.5	$29,400
2	Seal building air leaks	$1000	$5000	2 months	5	$19,000
3	Repair compressed air system leaks	$560	$800	1 month	1.5	$12,640
3	Replace exhaust fan motor (resizing)	$100	$500	2 weeks	5	$1900
4	Turn off all office computers during nonworking hours	$75	$0	1 week	0	$1800

Figure 11-4. The summary portion of an audit report provides the basic information about each of the recommended projects.

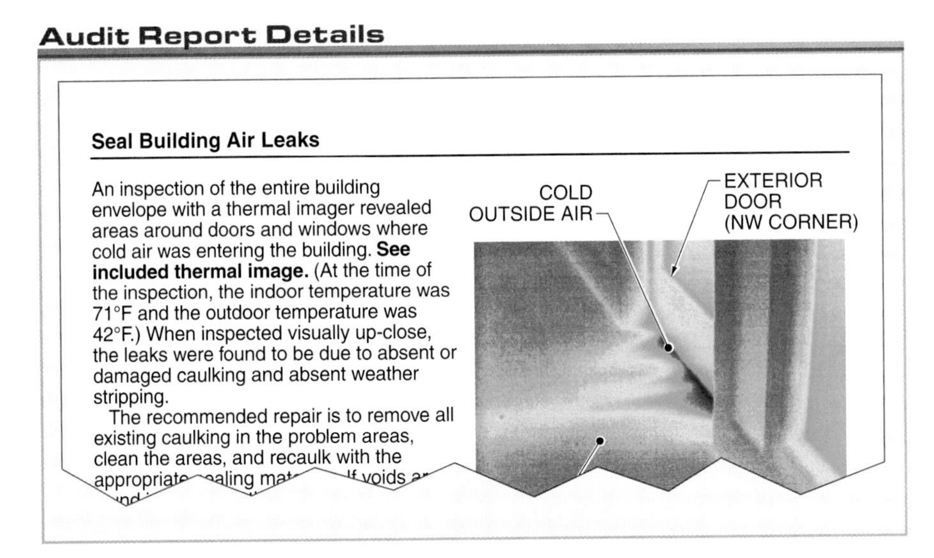

Figure 11-5. Energy audit findings are supported by the detailed descriptions of each savings opportunity.

For each savings opportunity, there may be more than one practical remedy. If multiple projects for a particular cause of waste are listed in the summary, they should all be described in the details section. Additional information should include the relative advantages and disadvantages of each choice.

IMPLEMENTATION

The audit report prioritizes savings opportunities because not all will be implemented, at least not initially. The projects with the highest ROI are typically chosen, while the others are kept for later consideration. Therefore, while the audit report should include estimated schedules, the final implementation plan cannot be compiled until the project list is determined.

When the initial projects are chosen, the audit team meets again to develop the implementation plan, which includes the action plan and project schedule. **See Figure 11-6.** The action plan lists the tasks to be completed, such as requesting bids and ordering equipment. Each task is assigned to an implementation team, with an audit team member being responsible for overseeing the progress. The original estimated schedule is then adjusted as needed to account for the actual task list.

As projects are being implemented, which will likely be over the span of a few months, the audit team should meet on a regular basis to track progress and verify the energy savings results. Audit team meetings are important for identifying projects that are not proceeding as planned and modifying the action plan and schedule to address problems.

Implementation Plans

Project	Task	Responsible Person	Other Team Members	Due Date
Seal building air leaks	Confirm and mark all leaks identified in audit	B. Smith	2 maintenance technicians	Apr 23
	Remove old caulking			Apr 30
	Clean and inspect transitions			May 5
	Apply new caulking			May 12
	Reinspect areas with thermal imager			May 15
Add VFDs to air handler fans and recalibrate controls f	Consult with HVAC technician to confirm project plans	J. Welch	HVAC technician	Apr 9
	Order VFDs		none	Apr 16
	Plan VFD installation loca		electrician	Apr 20

Figure 11-6. Once it has been decided which projects will be implemented, the audit team must develop a detailed plan and schedule for completing projects.

VERIFYING RESULTS

Audit team meetings are also useful for evaluating the tasks that have been completed and how they have affected energy use. The analysis of changes to utility bill charges can begin as soon as implementation of some projects has started and before all are completed. With the experience gained in identifying and analyzing utility charges, the verification process should be easier.

If savings are less than expected, the reason for the discrepancy must be determined. It may be necessary to repeat certain measurements in order to determine how the flow of power through the systems has changed and identify where the higher-than-expected consumption is. If the assumptions used for estimates were inaccurate, they must be reevaluated for future projects. Calculation errors can be avoided in the future by having multiple team members check the results.

Also, improvements to one area of a system can cause problems in other areas or even other systems. For example, adding more electronic controls and variable-frequency drives can cause power quality problems that harm other equipment. Comprehensive data logging and analysis is performed when all initial implementation projects are complete and accounts for any effects of one system on another. This information is compared with the baseline data and utility billing to verify the expected energy savings.

TECH-TIP

The most successful energy audits involve a variety of facility personnel, including those maintaining the equipment, reviewing the bills, and managing the tasks. Each contributor approaches an energy audit from a different perspective, which helps produce a comprehensive audit.

SUSTAINING EFFICIENCY

The energy auditing and efficiency improvement projects are usually significant investments in time and money. Even after the energy use of a facility has been reduced and the costs have been recouped, it is important to maintain an awareness of additional potential improvements. The experienced audit team should continue working to sustain the achieved results and identify additional opportunities. Once the initial energy audit is complete, subsequent follow-up work or repeat audits are more easily completed.

Periodic Inspections and Measurements

Even when a system has been fully inspected and adjusted for optimal efficiency, it should be regularly monitored for new problems. Equipment gradually ages, and even if a unit was operating acceptably during the previous audit, its condition may later deteriorate to a point where replacement is recommended. Therefore, major equipment and motors should be inspected periodically.

An inspection schedule should be implemented so that problems can be found quickly and equipment can be repaired before significant energy is wasted. Many of the inspection types included in the energy audit, such as thermal imaging inspections of various systems, may become part of the regular maintenance program. **See Figure 11-7.**

Continuous energy-use monitoring may be beneficial for some systems, such as certain large electrical loads. Data logging meters can be used on an on-going basis to log the power and energy parameters. Alternatively, electrical sensors can be permanently installed into panels or enclosures. The data from these sensors is sent to a computer system via a communication network.

Periodic Inspections

Fluke Corporation

Figure 11-7. Follow-up work after completing an energy audit and energy efficiency projects should include periodic inspections and measurements.

Preventive Maintenance

Preventive maintenance is the scheduled work required to keep equipment in peak operating condition. Preventive maintenance includes tasks such as inspection, cleaning, tightening, lubricating, and replacing consumable parts. Well-maintained equipment provides high-quality output at maximum efficiency and preserves safe conditions, while experiencing fewer malfunctions and failures. This results in increased service life, reduced downtime, and greater overall plant efficiency.

Even without any other energy-saving measures, improving preventive maintenance activities alone typically yields a very favorable ROI.

After completing an energy audit and energy efficiency program, it is critical for facilities to improve their preventive maintenance programs. **See Figure 11-8.** This maximizes the benefits of the changes implemented. Neglecting the maintenance program will likely prolong the payback period or even reverse the efficiency gains achieved. Also, a preventive maintenance program may also help the audit team to identify new or improved savings opportunities early. Therefore, the costs of implementing a remedy may be lower.

Fluke Corporation

Preventive maintenance involves an ongoing program of periodic inspections and maintenance work in order to keep equipment in peak operating condition.

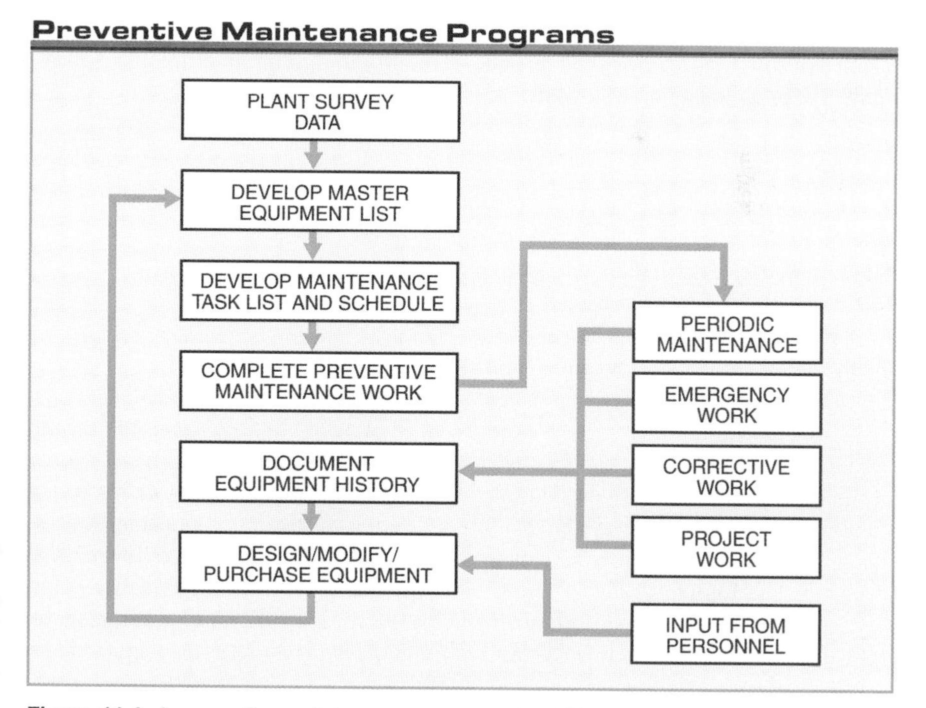

Figure 11-8. A preventive maintenance program should be continually improved through the knowledge and experience of the maintenance personnel.

The specific preventive maintenance tasks, and their optimal frequency to be performed, for each piece of equipment are determined by manufacturer specifications, equipment manuals, trade publications, and worker experience. These tasks may also require adjustments for changing conditions. For example, air handler filters are usually changed at certain intervals. However, extra dust in the air due to construction, dry and windy weather, or even a reorganization of workspaces can clog filters more quickly than normal. A proactive preventive maintenance program anticipates potential problems and monitors equipment more closely for increased maintenance requirements.

Improvements to the preventive maintenance program may require more personnel, better diagnostic equipment, more training, and improved procedures, all of which involve an additional investment. Furthermore, even though new equipment may have replaced old equipment with high breakdown maintenance costs, the new equipment must be adequately maintained or its efficiency will soon fall. These ongoing costs should be estimated and factored into the ROI calculations.

Facility Changes

An energy audit optimizes the operation of equipment for the current needs of a facility. If the facility changes the way it functions or modifies a system, the equipment may no longer provide the most efficient operations. Therefore, any permanent changes to the facility should be evaluated for any resulting impact on efficiency. A new energy audit may be necessary to return the facility to maximum efficiency following these changes.

Index

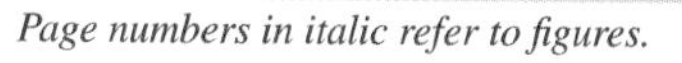

Page numbers in italic refer to figures.

G

H

I

L

M

N

O

P

R

S

T

U

V

W

Z

USING THE ENERGY AUDITING FOR INDUSTRIAL FACILITIES INTERACTIVE CD-ROM

Before removing the Interactive CD-ROM from the protective sleeve, please note that the book cannot be returned for refund or credit if the CD-ROM sleeve seal is broken.

System Requirements

To use this Windows®-compatible CD-ROM, your computer must meet the following minimum system requirements:

- Microsoft® Windows® 7, Windows Vista®, or Windows® XP operating system
- Intel® 1.3 GHz processor (or equivalent)
- 128 MB of available RAM (256 MB recommended)
- 335 MB of available hard disk space
- 1024 × 768 monitor resolution
- CD-ROM drive (or equivalent optical drive)
- Sound output capability and speakers
- Microsoft® Internet Explorer® 6.0 or Firefox® 2.0 web browser
- Active Internet connection required for Internet links

Opening Files

Insert the Interactive CD-ROM into the computer CD-ROM drive. Within a few seconds, the home screen will be displayed allowing access to all features of the CD-ROM. Information about the usage of the CD-ROM can be accessed by clicking on Using This Interactive CD-ROM. The Quick Quizzes®, Illustrated Glossary, Flash Cards, Fluke Virtual Meters, Reports and Forms, Media Clips, and ATPeResources.com can be accessed by clicking on the appropriate button on the home screen. Clicking on the American Tech web site button (www.go2atp.com) accesses information on related educational products. Unauthorized reproduction of the material on this CD-ROM is strictly prohibited.